Water Resource Economics

M.G. Chandrakanth

Water Resource Economics

Towards a Sustainable Use of Water for Irrigation in India

Springer

M.G. Chandrakanth
Department of Agricultural Economics
University of Agricultural Sciences
GKVK campus
Bangalore
India

ISBN 978-81-322-2478-5 ISBN 978-81-322-2479-2 (eBook)
DOI 10.1007/978-81-322-2479-2

Library of Congress Control Number: 2015940320

Springer New Delhi Heidelberg New York Dordrecht London

Printed on acid-free paper

Springer (India) Pvt. Ltd. is part of Springer Science+Business Media (www.springer.com)

This book is useful to students of agriculture, horticulture, forestry, fishery, sericulture and science; researchers and policymakers concerned with natural resource economics in general and water resource economics in particular. Dedicated to the lotus feet of my beloved and revered mother Smt V Gowramma and father Sri Gangadharaiah who toiled their life for my upbringing facing all odds.

Foreword

In the old days, it was common to tell a spendthrift not to spend money like water. The time has come, long since, to reverse this advice and tell people not to spend water like money! Water has certainly become much more precious than money, though those who are lucky enough to have an access to the resource easily enough to use it wastefully do not realise it. It is, therefore, important to spread awareness about the scarcity value of water, not only among end users like consumers but also among those who use it as an input in production, and among policymakers, planners and administrators. In this task, Prof MG Chandrakanth's enlightening book, *Water Resource Economics,* makes a valuable contribution. It may be felt that he has addressed the book more to fellow academicians and technical experts involved in policy-making and planning than to common people. It is also necessary, however, to educate and convince the former group first so that the right policies are formulated, and correct perceptions about the scarcity of water percolate to the common people effectively.

Though an ancient verse in Sanskrit describes the Earth as clothed in oceans (*Samudra-vasane Prithvee*), much of the seemingly abundant water is saline, and cannot be directly used, though of course it is because of these vast oceans that we get fresh water in the form of rains. As Chandrakanth points out, only 2.7 % is fresh or sweet water which is potable. The bulk of this—over 75 %—is frozen in the polar region, only 2.2 % is in the lakes, rivers and atmosphere, and nearly 23 % is groundwater (GW). Thus, much of the exploitable sweet water is in the ground. This is the reason why the main focus of the book is on GW. Sadly, this is not at all a renewable resource, since only about 10 % is rechargeable, and 90 % is exhaustible, that is, subject to permanent depletion. Sadly again, this exhaustible resource is not considered or treated as a common pool resource, and there is no management to guarantee its fair and equitable distribution, and also no system to ensure its sustainable use over time. Not only are different areas not endowed with an equal share of groundwater but also those who are lucky enough to strike abundant water consider it as their private resource with exclusive property rights, and there is hardly any check on the rate of withdrawal of GW to ensure equitable and sustainable use. This indeed is the crux of the GW problem, which is both economic and institutional in nature, and competently analysed by Chandrakanth in this book. He treats it mainly as a problem of externality, because the person who strikes GW and uses it in his land takes note of only the direct cost of extraction which governs his economic behaviour, but is indifferent to the cost arising on account of the deprivation of it to his or her neighbour and the cost of depletion involved.

The problem takes a different turn when it is not merely the individuals, but business enterprises which enter the scene as exploiters of groundwater for profit. For example, there were serious allegations against the Coca-Cola Company about overexploitation of GW in a village in Kerala to the detriment of the local population, who were deprived of drinking water significantly. The local people complained to the State Pollution Control Board and took the matter up to the Supreme Court, being unsatisfied with what the company offered as a compromise solution[1]. Chandrakanth has taken the case of sand mining on river beds, analysing its adverse impact on water availability (in Chap. 4 of the book). Sand mining beyond a depth of 2–3 ft seriously affects recharge of groundwater, making it even more exhaustible than it already is.

The problem is who will compensate, and by how much for causing such negative externalities. In the case of business enterprises, the solution appears simple as the liability rests on the enterprise which causes the deprivation. In the case of millions of farmers spread over the vast countryside causing these problems to their neighbours, the problem is institutionally much more difficult. It is a moot point how a farmer who strikes water in his borewell and makes it unavailable in his neighbour's borewell can be asked to compensate his neighbour, and by how much. There is also no law for it. What should be the permissible rate of extraction of water, and how to enforce it? We go beyond narrow microeconomics and enter into the complex realm of institutional economics here.

Chandrakanth rightly explores both the supply and demand side of the problem. It is not enough to merely augment supply, particularly while dealing with an exhaustible resource like GW; we cannot merely go for more borewells in areas where the resource is nearing exhaustion or is already exhausted. We have to seek different ways of moderating demand, both by institutional and technological means. By raising marginal productivity of water per unit through drip irrigation, as for example discussed by the author, substantial water can be saved and made available for distribution over wider areas. But how do we induce a farmer to adopt a relatively expensive technology like drip irrigation to save water and make it available to others? Chandrakanth discusses water markets as a possible solution (in Chap. 11) as they provide an incentive to both the sellers and buyers of water, and finds their efficacy in saving water and distributional implications positive and favourable. He also discusses the potential role of direct borewell recharge in sustaining GW in deep borewells. Since the size of the average agricultural holding has been rapidly declining in India over the decades, there will be greater scope for institutional means of sharing water. But a dilemma could arise here. Irrigation economics insists on maximising net returns per rupee cost of water, but a farmer

[1] For a brief account of the story, see M V Nadkarni (2014), Ethics for Our Times: Essays in Gandhian Perspective, New Delhi: Oxford University Press, pp. 255–56.

who thinks that his or her land is the scarcest resource will think of maximising net returns per unit of land. Of course, he may be willing to increase the net returns per unit cost of water also, in so far as it helps in raising the overall net returns from land, but not otherwise. If it pays him more to grow water-guzzling crops like paddy and sugarcane than to sell surplus water to his neighbour, he will certainly do so, unless he is restrained by law. From the social or national angle, however, saving water use in irrigation is a matter of great priority, since the bulk of GW and even surface water is used for irrigation. It is necessary to address the possible divergence between social and private interests by incentivising the adoption of social goals, including penalising the violation of social goals. But this also needs the creation of a regulatory and legal framework and its implementation, an area in which India is relatively weak.

The supply side of the problem also continues to be relevant, as Chandrakanth shows (in Chap. 10). He discusses here the economics of watershed development programmes, particularly as in Karnataka for illustration. These programmes have contributed significantly to improving the rate of recharge and renewability of GW and reduced the extent of negative externality. Pricing the water augmented is a tricky issue in a social cost–benefit analysis. Its shadow price should properly reflect the social value accorded to water.

Though the main focus of the book is on GW, it is not confined to it. The issue of GW use arises in the command areas of tank and canal irrigation also, and therefore, the book takes up the question of sustainable path of extraction of GW in such command areas. The author recommends an integrated approach to water management of multiple irrigation sources.

Chandrakanth gives special attention to the need for incentivising conservation of GW resources. It amounts to providing an ecosystem service, for which the people responsible should be rewarded, so that they are not tempted to either squander the resource or submit to the temptations offered by business enterprises interested in exploiting it. It is a tricky issue to decide the rate of payment for such services, as it has to be, on the one hand, effective and on the other hand, affordable for the economy. There is also the question of why a business enterprise should be denied access to the resource if it convincingly guarantees a sustainable use of it and creates a significant increase in local employment, contributing to the development of the region. It is, however, a delicate task to arrive at a solution balancing all these considerations in national as well as local interests. Sadly, ecological problems take place outside the market framework, but the solutions are sought only within the market framework. It is the burden of ecological economists like Chandrakanth to resolve this paradox, and he has shown considerable proficiency in doing so.

The book has succeeded in laying bare and solving quite a few complex riddles of water resource economics, with a focus on GW. It reflects the expertise of the author gained through decades of research into the problem, including honest and painstaking fieldwork. A special contribution of the book lies in providing empirical estimates of, for example, the variable cost and fixed cost of GW, in the case of negative externality involved. This is of great methodological interest. It comes as a welcome bonus, in addition to insights from theoretical and conceptual analysis.

I wish the book all success in disseminating expert knowledge of the field, which should be useful not only to university and research students but also to policymakers, planners and non-governmental organisations (NGOs) engaged in agricultural and rural development. Above all, I trust that the book succeeds in its chosen task of spreading awareness about the scarcity of GW and the need to ensure its sustainable use through a right mix of policy measures before it is too late.

July 11, 2015
Bengaluru

MV Nadkarni
Honorary Visiting Professor
Institute for Social and Economic Change
Bengaluru, 560072
(mvnadkarni1968@gmail.com)
(website: http://www.mvnadkarni.com).

Message

I have immense pleasure in writing a message for this book as the Vice Chancellor of the University of Agricultural Sciences, Bangalore (UASB) and as a colleague of Dr Chandrakanth. Currently, as the senior most professor in the UAS, Dr Chandrakanth has served as the Dean of the Agriculture College and is currently serving as a Professor and University Head of Agricultural economics. He took up research projects funded by the Ford Foundation in 1995 and developed the students and faculty in the field of natural resource and environmental economics with the World Bank National Agricultural Technology Project (NATP) funded 'Team of Excellence in Natural Resource Economics (TOENRE)' from 2000 to 2005. As the student exchange program coordinator, he has facilitated more than thirty students of the UASB to obtain an International Masters in Rural Development at the University of Ghent, Belgium with an Erasmus Mundus scholarship and many of them are pursuing doctoral studies in Germany.

India's dry agroclimatic zones in hard rock areas are fraught with severe biotic and abiotic stresses, and in addition, exposed to frequent droughts coupled with high GW extraction. Here, irrigation wells dug during the 1970s functioned till the mid-1970s. However, they could not sustain, as farmers deepened them further as dug-cum-borewells which did not function till the 1980s. Later, farmers began drilling shallow tubewells till 2005 which too did not function beyond 5–6 years. Since 2005, farmers began drilling deep borewells beyond 1500–2000 ft. Correspondingly, the extraction devices also have undergone change: from Yetha/Kapile/ Picota/ to centrifugal pumps and to heavy capacity submersible pumpsets. Thus, the extent of overexploitation by farmers has surpassed the limits of sustainability and resilience. Picking up a crucial contribution of this book, the cost of cultivation of GW irrigated crops does not include the component of cost of irrigation—the cost of GW. This underestimates costs and overestimates net returns to farmers.

This book is therefore topical, and has potential to serve as a textbook for students at undergraduate and postgraduate levels, faculty, researchers and policymakers concerned with natural resource and environmental economics. It provides valuable insights pertaining to the valuation of externality, policy, benefits of GW recharge and benefits of sharing water as well as demonstrated application of optimal control theory in GW management. This will be a valuable contribution to water resource economics and of great relevance to India, as most textbooks in natural resource and environmental economics have focused on issues outside India. I congratulate Dr Chandrakanth for this feat and wish him such endeavours.

Bangalore
9.7.2015

(H SHIVANNA)
Vice-Chancellor
University of Agricultural Sciences,
GKVK, Bangalore

Preface

Working on the Shadows

Water is one among the most important natural resources under threat that attracted large academic attention compared to its peers. Natural scientists as well as social scientists are vehemently involved in disentangling the complex issues (scientific/social scientific) pertaining to water and its uses. Economic, social, institutional and political solutions have been tried and failed at the altar and the problem continues to demand solutions. The world's largest share of usable water resource is used for irrigation and agriculture purposes, and overuse has led to many controversies and discussions.

The first to step in the arena of discussions were economists with a heroic idea of accurately pricing water resources, probably knowing quite well about the fluidity of the resource as also the changeability of pricing principles across the length and breadth of the use of the resource. Among social scientists, sociologists and political scientists were late in entering the fray and successfully opened a few important issues that additionally complicated the state of affairs. Further, when the environmental analysts entered the scrimmage on issues in the abuse of water and its impact on the environment, a sensitive disquietude began. The value of water (I have used the word 'value' here with a purpose) and the changes in it with every minute amends in micro situations pose difficult challenges. True, these are difficult to comprehend, but an analysis of the issues enveloping economics of water resources is a huge, complex but doable task. This text provides some important glimpses and earnest field experience about the basic economic issues confronting the water sector.

Among various sources of water, GW is the most widely used resource for irrigation in India as also in many parts of the world. It is dynamic in nature and the share rechargeable depends upon the type of aquifer. For instance, in hard rock aquifers, the share is abysmally low (5–10% of the rainfall). Thus, in relation to extraction, one should not overlook the fast depleting trapped GW resource in the aquifer. The very high rates of exploitation of GW are not deniable, and the externality that such exploitation exerts is environmentally extremely sensitive. The indiscriminate use of GW is mainly due to the ill-founded perceptions of its unending supply, ownership rights, irresponsibility towards its recharge, least respect for the interference caused by one well owner to another and above all, the sensitive negative externalities. More than the individuals who indiscriminately extracted GW, the state policies have encouraged such processes unabatedly. It is quite unfortunate that policy analysis and the role of the state in enhancing the use of GW (availability of easy finance National Bank for Agriculture and Rural Development, NABARD, different schemes allowing farmers to dig/bore a well) and disregard to the visible effects over the years has received marginal attention, but that is the job of analysts looking into state policies.

Water policies and GW regulations have been written and documented in India several times. But neither did they have the required gnash and teeth to get firmly on the ground, nor did these laws have any serious take on the environmental externalities. The extraction of GW went on unabated under the eyes of many ineffective regulations. As a result, failure of wells did not remain a freak instance, but became almost a high-probability event. The GW table was sinking to newer depths, and cumulative well interference became a nightmare for farmers and policymakers. The high probability of failure of wells inflicted high investment costs, huge outstanding loans, shrinking income sources and as a result, increasing acute farm sector distress. The failed well subsidy added to the already burgeoning problem by adding some more wells that have not even failed. Well interference, however, is a reality, and as a concept, it comes via cumulative seismic interference. The measurement of cumulative well interference is a challenge, and researchers have to depend on ground conditions. They need to rely on the reduction in GW levels in an extraction unit in response to the draft of GW from another unit of water extraction in the vicinity.

There are, of course, scientific experiments that employ stable radioisotopes like Tritium $^{3}H/^{14}C/HD^{16}O/H_2{}^{18}O/^{82}$ Br/$^{198/}$Au which are utilised to locate origin, age, velocity, direction of flow, interrelation between surface and GW or aquifers and the properties of aquifers. The experiments and the material are economically prohibitive in a situation like India, where the well density is very high in some regions and the aquifers have typical compositions. More than that, mere identification of interference does not automatically lead us to any solution of the problem and identification alone does not suffice. The GW tracing experiments can only enhance (sophisticate) the micro-information, which is otherwise easily available through the large density of failed wells. The actual solution lies in the institutional organisation of the users, water literacy, increasing recharge structure and controlled use of GW to keep draft/recharge ratio fairly in the neighbourhood of 1 but always less than 1.

Recharge of GW is the most critical aspect of GW economy, and it decides all the crucial parameters. Still, it has never been taken seriously in any policy document in India. There cannot be two opinions that we are fast losing the recharge zones due to deforestation, urbanisation, infrastructure builds and disappearing small water bodies (tanks). Percolation tanks came in a good policy intervention, but that was soon shelved like other water-conserving interventions. Rejuvenation of the old tanks guzzled a good amount of funds but with little sustenance, and tanks are disappearing very fast. Surprisingly, new problems like continuous sand extraction from river beds have emerged as an issue restricting even the present recharge flows. Sand mining is one of the best examples of this. Deepening of a river bed beyond a certain depth (deeper than 3 ft) leads to stoppage of river water recharging the GW by percolation through the banks. The danger of drying wells in the close vicinity of rivers is now a threatening situation. State policy towards sand extraction has not been effective due to unwilling implementation.

Enhancing recharge of GW through watershed development (WSD) is another intervention that yielded some good results. This policy tool is economically viable and environmentally acceptable. In India, we have a history of many WSD programmes (starting from early seventies to date) and the guidelines issued by the governments (both state and central). Each one of these guidelines makes some cosmetic changes, but as a result distorts the focus. It is a pity that there are hardly any remnants of the highly acclaimed World Bank-led WSD in Purua Nala, Maheshwar Nala, Manoli or in Kabbal Nala. Has the precious investment gone down the drain and has it delivered what it had promised? This irritating question can be asked even to the components that were taken in a missionary mode, for example, vegetative bunds through *Khus* grass planting. Not even a blade of that grass is visible on the locales for which crores of investment was put in the ground. It calls for accountability of policy that was applied without thinking. A systematic long-term sustained agenda would have made a sea change by now. But not everything in WSD is gloomy, and there are shining good examples wherein some good results are obtained. Economics of its sustenance and sociology of it in the institutional context is something that we look forward to.

The costing and valuation of water under different regimes and uses is a challenging task as the situations are dynamic, as are their interfaces with the sociopolitical determinants that complicate the computations. Something like electricity pricing, or for that matter marketing policy, can influence water pricing. After all, water is a natural resource used as an input and it defies all the rules of economics when it comes to pricing. No general principles could be used and one has to suffice with computations in a given precise situation or with very broad generalisation. This is also true about the estimates of its supply and demand, but in order to bring in discipline of use and inculcate water literacy, working even on shadows yields some results.

The technology of creating access to water, and markets that respond to the products achieved (using water as an input) decide the marginal value productivity. It is necessary to hasten and add that the environmental externalities should be essentially incorporated in any such computations. Therefore, the recent experiments on micro-irrigation have been, really, one decisive step towards obtaining a lasting solution. Drip irrigation is one such experiment that has proved economically viable and environmentally beneficial in many places. Initially, during the late eighties, even the agricultural scientists and policymakers were not favourably placed towards adopting this technology. In fact, some of them vehemently opposed any such intervention as it will gobble huge investments and may not sustain. However, time has proved that wrong and micro-irrigation emerged as one of the main leads that would ease the complex situation in the use of water, but that is not an endgame.

Attempting a book which encompasses many issues pertaining to economics of water resource inside two covers is a *Bhagirath* task. Water is in a liquid state and is an essential ingredient for survival. It is called by a synonym for life, *Jeevan*; dehydration of the human body is the finality of life. Similarly, being a *liquid,* it uniformly pervades the economic, social and political life of humanity. It is inseparable by the importance of its influence and hence, it encompasses all the social sciences. Institutions governing water use and enhancing water literacy assume greater importance as do state policy and interventions. The response of the state polity, policy and users has to be well-grounded in sustaining the use of water for agriculture and for other purposes. Water literacy hence becomes an important intervention.

Date: 7th July 2015

R S Deshpande
Former Director Institute for
Social and Economic Change
ICSSR National Fellow
Bangalore-560072

Acknowledgements

I am grateful to Prof. R. S. Deshpande, ex-Director, Institute for Social and Economic Change (ISEC), Bangalore (India), for initiating me to research on economics of groundwater through the NABARD-funded research project way back in 1982. It was in 1987 when as a Ford Foundation fellow at the University of California, Berkeley I had an opportunity to write a proposal on groundwater districts using Ciriacy-Wantrup's institutional economics framework under the guidance of Professor Jeff Romm of the College of Forestry. The proposal won me the Sigfried von Ciriacy-Wantrup postdoctoral fellowship in natural resource economics at the University of California, Berkeley for the years 1988–1990. Professor Jeff Romm, my mentor, gave me the contact of the previous Wanrup fellow, Dr. Brian Baker, who introduced me to Wantrupian Institutional Economics which immensely helped me to write the proposal on groundwater districts as a solution to address the groundwater predicament of India. I shared the draft proposal with Timothy Guinane, Peter Daugherty, George Frisvold, Lini Wollenberg, Nick Menzies, Yvonne Everett, Yeshwant Negi and Professors Jeff Romm, Keith Gilless, Louise Fortmann, William McKillop, Richard Norgaard, and Michael Hannemann for comments and suggestions which enriched my post-doctoral proposal and won me the Wantrup scholarship. Here, I worked with Professor Jeff Romm, Professor of forest resource policy, Professor William McKillop, Professor Keith Gilless, Professor Louise Fortmann, and Professor Richard Norgaard and from them I learnt forest economics, forest policy and institutional economics related to water resources. I am thankful to Lini/Eva Wollenberg, Kailash Govil, and other members of the Friday social forestry group at Mulford.

I must place on record the academic and moral support provided by my beloved teacher Professor R. Ramanna, the then Dean and Acting Vice Chancellor of the University of Agricultural Sciences, Bangalore (UASB) for deputing me to University of California, Berkeley on a Ford Foundation fellowship in 1987. Professor Ramanna along with teachers Professor S. Bisaliah, Professor G.S. Chandrashekar, and my sister MG Padma motivated me to pursue Masters in Agricultural Economics at UASB. I cannot forget my Master's thesis chairman Late Professor Noel Salvadore Paul Rebellow who taught me production economics, analytical and writing skills. Teachers Professors H.S. Krishnaswamy, P.C. Sridharan, H.M. Thimmarayappa, H.M. Parvathamma of Mathematics Department, Professors N. Sunderraj, Gurumurthy, Ramachandra of Statistics Department, made everlasting contribution to my understanding of agricultural economics at UASB. My Ph.D. Chairman Prof. J.V. Venkataram trained me in agricultural finance and macro-economics. Other teachers Prof. K.G. Mallikarjunaiah, Prof. G.K. Hiremath, Prof. H.G. Shankaramurthy, Prof. S.R.S. Murthy, Prof. Ranganatha Sastry, and Prof. Gundu Rao too contributed to my understanding of the subject.

The first Ford Foundation grant of US$ 41,000 to UASB was made by Dr. John Ambler for the project "Equity Issues in Groundwater Development: An Institutional Analysis of Failed Wells in Karnataka' for 1995– 2000. The second grant of US$ 80,000 was made by Dr. Ujwal Pradhan of the Ford Foundation for the project 'Dynamics of Access to Water Resources—A Resource Economics Study of the Watershed Development Program in Karnataka' 'for the years 2000–2005. The Indian Council of Agricultural Research (ICAR) and NATP (National Agricultural Technology Project) granted Rs. 60 lakhs for the research project – Team of Excellence in Natural Resource Economics (TOENRE), recognizing me as the leader of the team. Professor MV Nadkarni of the Institute for Social and Economic Change, Bangalore, Professor K Palanisami of Tamilnadu Agricultural University and Sri SV Jagannathan, hydrogeologist of Central Groundwater Board, educated me regarding the nuances of estimating negative externalities in groundwater irrigation. My colleague Dr. N. Nagaraj gave a firm scholarly support for my research projects which paved the way for joint research papers on water economics at UASB. The details of research work and the papers are available at http://toenre.com/. I thank profusely Dr. Mahadev G. Bhat, Professor of Environmental Economics, Florida International University, Miami, who further clarified concepts of existence value and optimal control theory applications in natural resource economics. I thank Dr. Yasmin Bagewadi and Ms. Balambal G. Subramanian who joined the TOENRE and helped the students understand and appreciate optimal control theory. Professor Indukumar along with Professor HS Krishnaswamy helped me to understand Optimal Control Theory and applications to water resource.

All my graduate students did excellent fieldwork in different agroclimatic zones of Karnataka on natural resource and institutional economics of groundwater and deserve due acknowledgement. Students who encouraged me to write this book are HM Seema, KM Bruhan and Kiran Kumar R. Patil. I must also make a special mention of B Shivakumaraswamy my first student who toiled hard in the initial stages of the Ford Foundation project. I should also thank Sushma Adya, V.K. Arunakumara, K.K. Ananda, A. Ravishankar, who strengthened my research efforts in Natural Resource Economics. My colleagues, Professors Ranganath Mangalvedkar, PS Srikanthamurthy, BM Ramakrishna at UASB have been a constant source of inspiration to write the book for the benefit of students. Colleague, Professor TN Prakash, also was supportive of my work on groundwater economics.

I am beholden to my beloved mother Smt Gowramma, father Sri Gangadharaiah for all their sacrifices, affection, commitment, and blessings showered. They shaped my personality, my life, teaching me from scratch, the do's and don'ts in life, by being philosophical and being role models themselves. I profusely thank and remain grateful to them and the almighty responsible for making me, what I am today. I place on record my sincere gratitude to my eldest brother Professor M.G. Nagaraja who has been a constant source of support and encouragement and similarly my sister Smt MG Padma, who took the role of my mother in educating me after my mother and father retired. My Srimathi M Shamaladevi and my daughters Alakananda Kevin and Dr. Mandakini Veeranarendra have helped and supported me constantly, bearing all of my unsystematic life pattern and forgetfulness and deserve special mention. I am grateful to all of them.

I sincerely appreciate the patience, commitment, and interest exhibited by Springer's Senior Editor, Business and Economics/ Law/Statistics, Smt. Sagarika Ghosh, who constantly encouraged me to finalize the book.

M.G. Chandrakanth
Professor and University Head,
Department of Agricultural Economics,
Leader of Team of Excellence in Natural
Resource Economics, University of
Agricultural Sciences, Bangalore

About the Author

M.G. Chandrakanth is currently the Professor and University Head of Agricultural economics and the former Dean of the College of Agriculture, University of Agricultural Sciences, Bangalore (UASB), India. He completed a BSc (Agri) degree in 1974, MSc (Agri) in agricultural economics in 1976 under Prof Noel SP Rebellow and a PhD in agricultural economics in 1981 under Dr JV Venkataram from the UASB. He served the Institute for Social and Economic Change as an Assistant Professor in the Rural economics unit from 1982 to 1983 under Prof VM Rao. He joined the UASB as an associate professor in 1983. He has been teaching agricultural economics, production economics, mathematical economics, advanced microeconomics, institutional economics and history of economic thought, natural resource and environmental economics courses since then.

Dr Chandrakanth pursued postdoctoral studies with a fellowship from the Ford Foundation, USA in forest resource economics between 1987 and 1988. He was awarded the Ciriacy-Wantrup Postdoctoral Fellowship for the period of 1988–1990, in institutional economics of natural resources at the University of California (UC), Berkeley, USA and worked under Prof Jeff Romm, specializing in water resource economics and policy.

He was nominated vice president of the Indian Society of Agricultural Economics in 2000 and a member of the editorial board of the *Indian Journal of Agricultural Economics* in 2012. He facilitated a student exchange program with the University of Ghent with Prof Guido Van Huylenbroeck in rural development (IMRD) due to efforts of his student Shalet Verghese and was Erasmus Mundus visiting scholar, University of Ghent, Belgium in 2008; DAAD scholar at Humboldt University, Berlin 2010 and 2012. He is a member of the Water Policy Law Group of the University of South Australia, Adelaide, Australia.

Dr Chandrakanth was awarded a Fellow of the Indian Society of Agricultural Economics in 2013 and awarded the *Dr Ramesh Chandra Agrawal Award of excellence* for his outstanding contributions in agricultural economics in 2014. He has published in the *Natural Resources Journal, Natural Resources Forum, Ecological Economics, Water Policy, Environment and Development Economics, Economic and Political Weekly, Indian Journal of Agricultural Economics, The Mysore Journal of Agricultural Sciences and the Agricultural Economics Research Review* (http://www.toenre.com/). He handled projects funded by the Ford Foundation from 1995 to 2005 on groundwater (GW) resource economics and has been the leader of the Team of Excellence in Natural Resource Economics (TOENRE) since 2000. Dr Chandrakanth is a student and admirer of professors R Ramanna, S Bisaliah, HS Krishnaswamy, HG Shankaramurthy, NSP Rebellow, KN Ranganatha Sastry, MV Nadkarni, VM Rao, K Palanisami and RS Deshpande for their academic excellence and commitment to teaching and is attempting to tread his path on similar lines with his passion for teaching and research in agricultural economics.

Contents

1 Water for Irrigation: An Overview 1
1.1 Why Water Is Vital for Agriculture? 1
1.2 How Much Water Does the Earth Receive? 2
1.3 Lessons and the Purpose of This Book 3
1.3.1 How Water Is Treated in Resource Economics Perspective 4
1.4 Why Water Is Vital for India's Agriculture 6
1.5 Himalayas Are 'The Roof of the World' 7
1.6 Monsoons Are Decisive for India's Economy 7
1.7 The HRA and 60% Syndrome 7
1.8 Public Investment in Surface Water and Private Investment in Groundwater 8
1.9 What Is Well Interference and Cope of Depression and How it Affects Groundwater Extraction? 8
1.10 Consumptive Use 8
1.11 India—Top Extractor of Groundwater in the World 9
1.12 Carbon Emissions 10
1.13 Physical Properties of Surface Water of Relevance to Economics 10
1.14 Physical Properties of Groundwater of Relevance to Economics 11
1.15 Water as an Economic Good: Role of Literacy and Governance 11
1.16 Sustainability in Water Use Requires Sacrifice 11
1.17 What Is Sustainable Extraction of Groundwater 12
1.18 Property Rights to Groundwater with the Advent of Borewells 13
1.18.1 How Definite Are Property Rights to Groundwater? 13
1.19 Open Access Resource? 14
1.20 Common Property Resource and Reciprocal Externalities 15
1.21 'The Six Blind Men and an Elephant' 16
1.22 Markets, Technologies, Institutions, Social Capital, Transaction Costs and Governance Structures 16
1.23 Groundwater and Climate Change 17
1.24 Does Jevons Paradox Apply to Groundwater Extraction? 17
1.25 Supply Side Factors Governing Groundwater 18

1.26 Demand Side Factors Governing Groundwater ... 18
1.27 Wide Estimates of Water Extraction and Use of Electricity ... 18
1.28 Power Supply to Irrigation Pumpsets Is Treated as a Residual ... 19
1.29 Who Bears the Brunt of the Cost of Groundwater Despite Power Subsidy? ... 20
1.30 Water Organizations in India ... 21
1.30.1 Estimation of Reciprocal Negative Externalities ... 21
1.30.2 Water-Use Efficiency ... 23
References ... 23

2 Externality in Irrigation ... 25
2.1 Externality ... 25
2.1.1 Reciprocal Externality ... 25
2.1.2 Explanation of Externality ... 27
2.1.3 Why Farmers Overextract Groundwater: Perspective of Negative Externality ... 27
2.1.4 Why Farmers Hesitate to Invest in Groundwater Recharge: Perspective of Positive Externality ... 28
2.1.5 Interactive Effects Among Wells ... 30
2.1.6 Why Study Interference? ... 30
2.1.7 How to Study Interference ... 31
2.1.8 PRA Approach to Locate the Area of Cumulative Interference Within a Selected Village ... 32
2.1.9 PRA Mapping: Wealth of Results on Groundwater Wells, Interference ... 32
2.1.10 Dowsing ... 33
2.2 PRA Results ... 34
2.2.1 Winner's Curse ... 35
2.2.2 Role of Groundwater Management ... 35
References ... 36

3 Unidirectional and Reciprocal Externality in Irrigation ... 37
3.1 Preamble ... 37
3.2 Explanation ... 38
3.3 Cultivation of Water-Intensive Crops Enhances Negative Externality ... 40
3.4 The Measure of UDNE ... 41
References ... 42

4 Sand Mining Externality ... 43
4.1 Encroachment ... 43
4.2 Overlap of Externalities ... 44

4.3 Sand Mining in Palar Pennar Basin 44
4.4 Unidirectional Negative Externality Leading to Overextraction of Sand from River Streams 44
4.5 Unidirectional Externality 46
4.6 Reciprocal Externality 47
4.7 Amortization of Investment 47
4.8 Encroachment of Riverbeds and Failure of Filter Point Wells 47
4.9 Sample 48
4.10 Land Use Pattern 49
4.11 Influence of Sand Mining on Well Failure, Age and Life of Irrigation Wells 50
4.12 Cropping Pattern and Net Income 51
4.13 Net Returns per Acre under Different Well Regimes Considering the Proportion of Well Success and Average Age of Well 53
4.14 Sources of Income 55
4.15 Net Income per Acre from Different Activities 56
4.16 Access to Groundwater Resources 56
4.17 Economics of Irrigation Considering Distance from Stream 62
4.18 Details of Irrigation Wells Before and After Mining in Sand Mining Areas 63
4.19 Opportunity Cost of Providing Agricultural Land for Transporting Sand 64
4.20 Analysis of Sand Mining Accounting for Externalities 65
4.21 Economics of Sand Transportation from Uttara Pinakini Riverbed in Gauribidanur Taluk to Bangalore Megacity 66
4.22 Cobb Douglas Function for Demand for Sand in Megacities 68
4.23 Conclusions 81
References 82

5 Relationship Between Rainfall and Recharge 83
Reference 85

6 Marginal Productivity of Water 87
6.1 Preamble 87
6.2 Average Product Influenced by Extreme Values 88
6.3 Marginal Productivity 89
6.4 Division of Labour Is Impossible in Farming 90
6.5 Observed Reality and Estimated Utopian 90
6.6 Interpreting Marginal and Average Products 91
6.7 Approach to Understanding the Marginal Product 91
6.8 The Case of Multiple Inputs 92
6.9 Understanding Ceteris Paribus or Keeping Other Things Constant 92

7 Costing Water for Irrigation 95
7.1 Properties of Water as a Natural Resource 95
7.2 Hitherto, Why Cost of Water/Groundwater Was Not Accounted in Irrigation 97
7.2.1 Surface Water 97
7.2.2 Groundwater 98
7.3 Concept of Initial Failure, Subsistence Life, Premature Failure and Economic Life of Wells 99
7.4 Costing Groundwater in Irrigation 100
7.4.1 Amortized Cost of Borewell 100
7.4.2 Yield of Irrigation Borewell 103
7.4.3 Groundwater Use in Conventional Irrigation System 103
7.4.4 Groundwater Use Measurement in Drip Irrigation System ... 103
7.5 Annual Cost of Irrigation 103
7.5.1 Externality Defined 104
7.5.2 Illustration of Cost of Groundwater Irrigation 104
7.6 Conclusions 107
References 107

8 Locating Interference and Valuing Water 109
8.1 Methodology to Identify/Locate Interference Among Irrigation Wells 110
8.2 Growth in Groundwater Investment 112
8.3 Dynamic Nature of Rights 113
8.4 Property Rights to Groundwater Are Obscure 113
8.5 Intertemporal Externality 114
8.6 Policy Governing Groundwater Irrigation 114
8.7 Relevance of Valuing Groundwater 115
8.8 Estimation of the Cost of Electrical Energy Used 116
8.9 Estimated Groundwater Values 117
8.10 Electricity Subsidy 117
8.11 Groundwater Markets Uncommon in Peninsular India 120
8.12 Value of Groundwater in Cost of Production 121
8.13 Variation in Net Returns and Electricity Used 121
8.14 Epilogue 122
References 123

9 Demand Side Economics of Micro-irrigation 125
9.1 Preamble 125
9.1.1 Status 125
9.1.2 Cropping Pattern 126
9.1.3 Well Failure 126
9.1.4 Profile of Irrigation Wells 129
9.1.5 Net Returns in Drip-Irrigated Farms and Conventional-Irrigated Farms 129

9.1.6 Technical Efficiency 132
9.1.7 Economic Efficiency of Water Use 132
9.2 Marginal Productivity of Groundwater in Drip and Conventional Irrigation Farms 132
9.3 Net Return Function for Conventional Irrigation Farms 135
9.4 Investment Behaviour of Drip Irrigation Farmers (Tobit) 135
9.5 Results of the Discriminant Function Analysis 136
9.6 Conclusions 137
References 138

10 Supply-Side Economic Contribution of Watershed Programme to Groundwater Recharge 139
10.1 Preamble 139
10.2 Methodology 139
10.2.1 Contribution of the Watershed Programme for Farmers Not Possessing Irrigation Wells 141
10.2.2 Contribution of the Watershed Programme for Farmers Possessing Irrigation Wells 143
10.2.3 Overall Contribution of the Watershed Programme for Farmers Dependent on Rainfall as well as for Farmers Possessing Irrigation Wells 143
10.3 Overall Contribution of Watershed and Rainfall 147
10.4 Implications of the Study 148
Reference 148

11 Water Markets for Sharing Limited Water 149
11.1 Groundwater Markets (GWMs) for Sharing Limited Water—A Recent Development 149
11.2 GWMs and Efficiency 150
11.3 Hypotheses 150
11.4 Sampling 150
11.5 Yield of Irrigation Well 158
11.6 Cost of Production 158
11.7 Economic Profile of Farmers Involved in Groundwater Market 159
11.8 Features of Irrigation Wells 160
11.9 Cropping Pattern 162
11.10 Sources of Income for Farmers in Groundwater Market 162
11.11 Economics of Irrigation 164
11.12 Income from Sale of Groundwater 164
11.13 Groundwater Sold for Different Purposes 167
11.14 Institutional Arrangements in Groundwater Market 168
11.15 Estimated Price of Groundwater in Nice Water Markets According to End Use 168
11.16 Nash Equilibrium Model of Groundwater Pricing 169
11.17 Implications 171
References 172

12 Sustainable Path of Extraction of Groundwater in Tank and Canal Command Areas ... 173
12.1 Model ... 173
12.2 Estimation of NB Under Myopic (No Control) ... 181
12.3 Intertemporal Allocation of Groundwater ... 182
12.4 Implications ... 189
References ... 189

13 Water Policy ... 191
13.1 Weak Points of Water Policy ... 193

14 Estimated Cost of Surface Water and Payment for Ecosystem Services (PES) for Water ... 197
14.1 Culture of Sacred Groves ... 197
14.1.1 How PES Is Linked with Farmers' Efforts in the Cauvery Basin ... 197
14.2 Cost of Water ... 198
14.3 PES for Water ... 198
14.4 Payment Vehicle ... 199
References ... 200

15 Economics of Artificial Recharge of Borewell in Hard-Rock Areas ... 201
15.1 Recharge of Borewells ... 201
15.2 Economic Benefit of Recharge Using Partial Budgeting ... 202
15.3 Added Costs ... 202
15.4 Reduced Returns ... 202
15.5 Reduced Costs ... 203
15.6 Added Returns ... 203
15.7 Conclusion ... 204
References ... 204

16 Economics of Sharing Irrigation Water ... 205
16.1 Subdivision and Fragmentation of Holdings ... 205
16.2 Groundwater Friendly Crop Pattern ... 205
16.3 Costs and Benefits of Sharing Water for Irrigation ... 206
16.4 Marginal Productivity of Groundwater Irrigation ... 206
16.5 Transaction Costs ... 207
References ... 208

Index ... 209

List of Abbreviations and Acronyms

ACAW	Amortized cost of all wells
ACFW	Amortized cost of functioning wells
AL	Average life/age of borewell
ANOVA	Analysis of variance
AP	Average product
BRS	Borewell recharge structure
BW	Borewell
BWFNA	Buyers of water for nonagricultural purpose
CACP	Commission for agricultural costs and prices
CD	Cobb–Douglas production function
CIF	Conventional irrigation farmers
CS	Conveyance structure
DAAD	German Academic Exchange Service
DIF	Drip irrigation farmers
DIS	Drip irrigation system
DPAP	Drought Prone Area Programme
EDZ	Eastern dry agroclimatic zone of Karnataka
FBWFA	Farmers buying water for agricultural purposes
FME	Farm management economics
FNSW	Farmers not selling water
FSWFA	Farmers selling water for agricultural purposes
FSWFNA	Farmers selling water for nonagricultural purposes
GCA	Gross cropped area
GHG	Green house gas emissions
GIA	Gross irrigated area
GPH	Gallons per hour
GUII	Groundwater use intensity index
GW	Groundwater
GWCI	Groundwater wells under canal irrigation command
GWM	Groundwater market
GWSI	Groundwater wells under sole irrigation (without any recharge)
GWTI	Groundwater wells under tank irrigation command

HDPE	High density polyethylene
HP	Horse power
HRA	Hard rock area
Hss	Steady-state height of the groundwater level
ICAR	Indian Council of Agricultural Research
IP set	Irrigation pump set
IWMI	International Water Management Institute
KWH	Kilowatt hour
LDMR	Law of diminishing marginal returns
LDMU	Law of diminishing marginal utility
ICRISAT	International Crops Research Institute for the Semi-Arid Tropics
ISI	Indian Standards Index
MC	Marginal cost
MCM	Million cubic meter
MIS	Micro-irrigation structure
MP	Marginal product, marginal productivity
MOEF	Ministry of Environment and Forests
MR	Marginal return
MSP	Minimum support price
MYO	Myopic extraction of groundwater
NASA	National Aeronautic Space Administration
NATP	National Agricultural Technology Project
NB	Net benefit
NBD	Negative binomial distribution
NC	Not cultivated
NSMA	Non-sand mining area
OC	Opportunity cost
OPT	Optimal extraction of groundwater using optimal control
OSS	Overground storage structure
OZ	Ozone
P and A	Pumpset and accessories
PBP	Payback period
PE	Production economics
PES	Payment for ecosystem services
PRA	Participatory rural appraisal
PVC	Polyvinyl chloride
PVNB	Present value of net benefits
RDPR	Rural Development and Panchayat Raj
RT	Record type
SANDEE	South Association Network for Development and Environmental Economics
SAS	Statistical Analysis System
SMA	Sand mining area
SVCW	SV Ciriacy-Wantrup
TB	Total benefit
TC	Total cost

TOENRE	Team of Excellence in Natural Resource Economics
TP	Total product
TR	Total return
TWU	Total water used
UASB	University of Agricultural Sciences, Bangalore
UDNE	Unidirectional negative externality
VDSA	Voluntary Data Sharing Agreement
WDP	Watershed Development Program or Project
WUF	Water used on the farm
Wss	Steady-state groundwater extraction

List of Tables

Table 1.1	Table key differences between economics, natural resource and environmental economics	5
Table 1.2	Organizations monitoring water resource in India	22
Table 3.1	Ratio variables and hypotheses concerning negative externality in well irrigation in hard rock areas (HRAs)	39
Table 3.2	Logit probability of drilling an additional well by farmers, given well failure	40
Table 3.3	Estimation of willingness to pay for an additional well (Tobit)	41
Table 4.1	Extent of sand mining along the rivers in Kolar District	45
Table 4.2	Depth of sand mining, sample size, in Pinakini basin in Gauribidanur, Karnataka	48
Table 4.3	Land use pattern and net income in sand mining and non-sand mining	49
Table 4.4	Age and life of irrigation wells in sand mining and non-sand mining areas of Gauribidanur taluk along Uttara Pinakini river, Kolar district, Karnataka	50
Table 4.5	Cropping pattern, area and net income under open well (near streams) in NSMAs and SMAs along the Uttara Pinakini river basin in Gauribidanur taluk, Karnataka	51
Table 4.6	Cropping pattern, area and net income under filter point well (near streams) in NSMAs and SMAs along Uttara Pinakini river basin in Gauribidanur taluk, Karnataka	52
Table 4.7	Cropping pattern, area and net income under borewell (located away from streams) in NSMAs and SMAs along the Uttara Pinakini river basin in Gauribidanur taluk, Karnataka	53
Table 4.8	Cropping pattern, area and net income from Agriculture in NSMAs and SMAs along the Uttara Pinakini river basin in Gauribidanur taluk, Karnataka	54

Table 4.9 Net returns per acre under different well regimes considering the proportion of well success and average age of well in SMAs and NSMAs along the Uttara Pinakini river basin in Gauribidanur taluk, Karnataka............... 54
Table 4.10 Percentage difference of net returns per acre considering the proportion of well failure and average age of well along the Uttara Pinakini river basin in Gauribidanur taluk, Karnataka 55
Table 4.11 Sources of income in SMAs and NSMAs along the Uttara Pinakini river basin in Gauribidanur taluk, Karnataka. (₹ per farm per year).. 55
Table 4.12 Net income per acre from different activities in SMAs along the Uttara Pinakini river basin in Gauribidanur taluk, Karnataka (₹ per acre per year)... 56
Table 4.13 Net income per acre from different activities in NSMAs along the Uttara Pinakini river basin in Gauribidanur taluk, Karnataka (₹ per acre per year)... 57
Table 4.14 Access to groundwater resources for irrigation in NSMAs and SMAs from *open wells* along the Uttara Pinakini river basin in Gauribidanur taluk, Karnataka 57
Table 4.15 Access to groundwater resource for irrigation in SMAs and NSMAs from *filter point well (FPWs)* along the Uttara Pinakini river basin in Gauribidanur taluk, Karnataka............... 59
Table 4.16 Access to groundwater resources for irrigation in SMAs and NSMAs from *borewells* along the Uttara Pinakini river basin in Gauribidanur taluk, Karnataka 60
Table 4.17 Access to groundwater resources for irrigation in NSMAs and SMAs from all wells along the Uttara Pinakini river basin in Gauribidanur taluk, Karnataka 61
Table 4.18 Economics of irrigation according to distance from river stream in SMAs and NSMAs along the Uttara Pinakini river basin in Gauribidanur taluk, Karnataka............................ 63
Table 4.19 Details of groundwater yield of irrigation wells before and after SMA along the Uttara Pinakini river basin........................ 64
Table 4.20 Opportunity cost of providing agricultural land for transporting sand.. 65
Table 4.21 Partial budgeting analysis of sand mining by farmers without considering externalities due to sand mining in Gauribidanur taluk, Karnataka.. 66
Table 4.22 Partial budgeting analysis of sand mining in the farm accounting for externalities along the Uttara Pinakini river in Gauribidanur taluk of Karnataka.. 67
Table 4.23 Economics of sand transportation from Uttara Pinakini riverbed in Gauribidanur taluk to Bangalore city 68

Table 4.24 Estimation of demand function for sand (volume in cubic feet) in the megacity of Bangalore, 2003 (Cobb Douglas function) 69
Table 4.25 Documentation of travails due to sand mining activity along the Uttara Pinakini river basin in Gauribidanur taluk, Karnataka 70
Table 4.26 Transitional probabilities of cropping pattern in Tharati village in Tumkur district of Karnataka (1998–2011) 76
Table 6.1 Marginal product (MP) in farm management economics (FME) and production economics (PE) 93
Table 7.1 Economics of vegetable crops including groundwater cost of irrigation on farms using drip irrigation in Karnataka 105
Table 7.2 Economics of perennial crops including groundwater cost of irrigation in Karnataka 106
Table 8.1 Net return considering value of groundwater and cost of electrical energy across agroclimatic zones of Karnataka 118
Table 8.2 Average electricity used and net returns realized by groundwater farmers in Karnataka 120
Table 9.1 Cropping pattern in drip and conventional irrigation farms in the eastern dry zone of Karnataka (acres) 127
Table 9.2 Distribution of irrigation wells across the size of holding in the eastern dry agroclimatic zone (EDZ) of Karnataka 128
Table 9.3 Profile of irrigation wells on drip and conventional irrigation farms 130
Table 9.4 Net returns of crops under drip and conventional irrigation 131
Table 9.5 Water use efficiency in drip and conventional irrigation farms in the eastern dry zone of Karnataka 133
Table 9.6 Investment behaviour by drip irrigation farms (Tobit model). Dependent variable: investment in rupees per farm on drip irrigation farms 135
Table 9.7 Factors discriminating drip and conventional irrigation farms in the eastern dry zone of Karnataka 136
Table 10.1 Estimated contribution of the Sujala watershed development programme exclusively for farmers who totally depend on rainfed agriculture (and do not possess irrigation wells) on the Veda River bank in Chitradurga district (net returns Rs. per acre) 142
Table 10.2 Estimated contribution of the Sujala watershed development programme exclusively for farmers who possess irrigation wells on the Veda River bank in Chitradurga district (₹ per acre) 143
Table 10.3 Estimated contribution of the Sujala watershed development programme on the Veda River bank in Chitradurga district (₹ per acre) 144

Table 10.4 Comparison of the net returns per acre of the Sujala watershed over the non-Sujala (DPAP) watershed and non-watershed area across different categories of farmers at the Veda River, Chitradurga district 145
Table 10.5 Incremental net returns due to the Sujala watershed over the non-Sujala watershed area and the non-watershed area on the Veda River bank in Chitradurga district in a drought year (2004) 146
Table 10.6 Estimated contribution due to the Sujala watershed development programme, institutional innovations, people's participation and rainfall in Chitradurga district (₹ per acre)...... 147
Table 11.1 Distribution of sample respondents in groundwater markets (GWMs) in Sidlaghatta taluk, Karnataka (2004)........ 151
Table 11.2 Socioeconomic features of groundwater seller and groundwater buyer farmers in Sidlaghatta taluk, Karnataka 160
Table 11.3 Particulars of irrigation wells of sample farmers in GWM in Sidlaghatta taluk, Karnataka (2003–2004)........ 161
Table 11.4 Cropping pattern of different categories of sample farms in GWM in Sidlaghatta taluk, Karnataka (2003–2004) (area in acres) 163
Table 11.5 Sources of farm income in different categories of sample farms in GWM in Sidlaghatta taluk, Karnataka (2003–2004) (net returns in ₹ per farm per year) 164
Table 11.6 Economics of irrigation in different categories of sample farms in GWM in Sidlaghatta taluk, Karnataka (2003–2004).... 165
Table 11.7 Net returns from agriculture and from sale of groundwater in GWM in Sidlaghatta taluk, Karnataka (2003–2004) 166
Table 11.8 Economy of groundwater buyers for irrigation in Sidlaghatta taluk, Karnataka (2003–2004) 167
Table 11.9 Details of groundwater sold for different purposes in GWM in Sidlaghatta taluk, Karnataka (2003–2004) 168
Table 11.10 Estimated price of groundwater in different crops in GWM, Sidlaghatta taluk, Karnataka (2003–2004) 169
Table 11.11 Price of groundwater purchased for nonagricultural purposes from farmers in GWM in urban fringe of Sidlaghatta town Karnataka (2003–2004)........ 170
Table 11.12 Nash equilibrium model of groundwater niche market dependent variable: Ratio of groundwater price to its amortized cost ($n=30$) 171
Table 12.1 Economic and hydrological parameters of the estimated optimal control model 181
Table 12.2 Estimates of quadratic water revenue function 182
Table 12.3 Features of irrigation wells in North Karnataka, Belgaum 183
Table 12.4 Myopic extraction of groundwater, pumping lift and PVNBs in GWSI 183

Table 12.5 Optimal extraction of groundwater, pumping lift and PVNB in GWSI 184
Table 12.6 Myopic extraction of groundwater, pumping lift and PVNB in GWTI 185
Table 12.7 Optimal extraction of groundwater, pumping lift and PVNB in GWTI 185
Table 12.8 Myopic extraction of groundwater, pumping lift and PVNB in GWCI 186
Table 12.9 Optimal extraction of groundwater, pumping lift and PVNB in GWCI 187
Table 15.1 Economic impact of artificial recharge of borewell (₹/borewell) 203
Table 16.1 Economics of farms sharing water for irrigation 206
Table 16.2 Gross returns per farm regressed on water used 207

List of Figures

Fig. 1.1 Well interference and cone of depression 8

Fig. 1.2 Water resource economics—heuristic net social benefits from groundwater use according to economics, natural resource economics, environmental economics and ecological economics 12

Fig. 2.1 Negative (reciprocal) externality leading to overextraction of groundwater 28

Fig. 2.2 Positive externality leading to low groundwater recharge effort 29

Fig. 4.1 Reciprocal negative externality due to sand mining in Uttara Pinakini river basin of Karnataka.................................... 46

Fig. 4.2 Farmers engaged in sand mining activity along the Uttara Pinakini river stream in Gowribidanur taluk, Karnataka, India, 2003, Pic: MG Chandrakanth and AC Hemalatha............. 71

Fig. 4.3 Even graveyards are not spared from sand mining activity in Gowribidanur taluk, Kolar district, Karnataka, India (by MG Chandrakanth and AC Hemalatha) 71

Fig. 4.4 Intense sand mining activity to a depth of 30 ft in Uttara Pinakini river stream in Gowribidanur taluk, Kolar district, Karnataka, India (by MG Chandrakanth and AC Hemalatha)...... 72

Fig. 4.5 Sand mining to a depth of 30 ft in Uttara Pinakini river stream in Gowribidnaur taluk (by MG Chandrakanth and AC Hemalatha), Kalludi in Gauribidanur Taluk, Karnataka, 2003 72

Fig. 4.6 Sand being washed using compressor pump to remove red sting as white coloured sand fetches high price, Gauribidanur taluk, Karnataka (by MG Chandrakanth and AC Hemalatha).................................... 73

Fig. 4.7 Sand being mined in agricultural field near the Uttara Pinakini river stream in Heribidanur village, Gauribidanur taluk, Karnataka, 2003 (by MG Chandrakanth and AC Hemalatha) 73

Fig. 4.8 Railway bridge foundation frequently being repaired due to sand mining across Uttara Pinakini river in Gauribidanur taluk, Karnataka (by MG Chandrakanth and AC Hemalatha) 74
Fig. 4.9 Public protesting against sand mining activity in Gauridanur town, 2002 74
Fig. 4.10 Tharati groundwater scenario 79
Fig. 4.11 Sweet flag crop of pre 2000 due to abundant water 80
Fig. 4.12 Arecanut replacing sweet flag in 2013 due to groundwater scarcity in Tharati 80
Fig. 4.13 Lone farm with sweet flag in 2013 in Tharati 80
Fig. 4.14 Sweet flag replaced by Chrysanthemum in 2013 81
Fig. 6.1 Stages of production as a function of groundwater irrigation 88
Fig. 9.1 Marginal productivity of groundwater influenced by technology (of drip irrigation). *DIF* drip irrigation farmers, *CIF* conventional irrigation farmers 134
Fig. 10.1 Boulder checks in coconut orchard, Sujala watershed, Karnataka 140
Fig. 10.2 Spillway constructed for runoff in the Sujala watershed, Karnataka 140
Fig. 10.3 Recharge pit for irrigation borewell in the Sujala watershed programme 141
Fig. 11.1 Over head tank constructed for filling groundwater to the tankers in water market by SC Deepak 152
Fig. 11.2 Bullock drawn tanker at a water filling station by SC Deepak 152
Fig. 11.3 Groundwater purchased for domestic purposes from bullock drawn water tankers by SC Deepak 153
Fig. 11.4 Beetroot crop grown in water market area by SC Deepak 153
Fig. 11.5 Tomato—a major vegetable crop grown in water market area by SC Deepak 154
Fig. 11.6 Ragi crop grown in irrigated condition in groundwater market (GWM) area by SC Deepak 154
Fig. 11.7 Mulberry—a major crop cultivated by water market farmers by SC Deepak 155
Fig. 11.8 Silkworm rearing by water market farmer in rearing house by SC Deepak 155
Fig. 11.9 Groundwater seller and buyer (*first* and *second* from *left*) in seller's field by SC Deepak 156
Fig. 11.10 Measuring the groundwater yield of the bore well in GWM by SC Deepak 156
Fig. 11.11 Power tiller drawn water tanker used for conveying-selling water to filatures by SC Deepak 157
Fig. 11.12 Silk filature utilizing groundwater for boiling silk cocoons by SC Deepak 157

Fig. 12.1 Farms cultivating **a** sunflower and **b** tomatoes under sole well situation and **c** paddy under tank command in Ramadurga taluk, Belgaum district 174
Fig. 12.2 Dug well in canal command, Saundatti taluk, Belgaum district 175
Fig. 12.3 Dug well with low water level in sole well situation, in Bailhongala taluk, Belgaum district 176
Fig. 12.4 Myopic and optimal groundwater extraction, pumping lift and present value net benefits *(PVNB)* in GWSI 177
Fig. 12.5 Myopic and optimal groundwater extraction, pumping lift and present value net benefits (*PVNB*) in GWTI 178
Fig. 12.6 Myopic and optimal water extraction, pumping lift and present value net benefits (*PVNB*) in GWCI 186
Fig. 12.7 Dug well with water level, indicating synergistic effect of canal in Ramdurga taluk, Belgaum district 188
Fig. 12.8 Failed open well in sole well situation, Bailhongala taluk, Belgaum district 188

Chapter 1
Water for Irrigation: An Overview

Water is life's mater, matrix, mother and medium. There is no life without water.
Water is the only drink for a wise man

1.1 Why Water Is Vital for Agriculture?

In India, food grains account 63 % of the cropped area, with 48 % of food grain area and 45 % of the cropped area being irrigated. Rice constitutes 42 % and wheat 35 % of food production, of which 59 % of rice and 91 % of wheat are from irrigated area contributing to food security, stability, productivity and resilience. At present, more than 60 % of food production is from irrigated areas in India, and more than 80 % of irrigation is from groundwater resource. As small and marginal holdings constitute around 83 % of operational holdings operating 41 % of the area, violation of isolation distance among wells is by rule than by exception as every farmer respects the Indian Easement Act (http://www.nih.ernet.in/rbis/rights.htm) where the land owner has legitimate right for the water under his/her land. Howeer when this Act was in operation in 1882, the science of hydrogeology perhaps did not explain the land owners that when ever any land owner is withdrawing water/groundwater, she/he in fact is drawing groundwater from neighbouring farmer's lands due to the interaction of cones of depression due to cumulative interference among irrigation wells, described with a diagram in the coming chapter. Hence, interference among irrigation wells is a crucial aspect in the economics of groundwater irrigation. Pumping in one well too much and too fast can dry up the neighbouring wells located in a common aquifer due to well interference. It is further difficult to delineate the aquifer

The purpose of this book is capacity building, highlighting the role of water resource economics as a process towards sustainable extraction and use for maximizing social welfare. This serves as a text book and a guidebook for users and policy-makers regarding the 'wise' use of scarce water resource in all its uses.

M.G. Chandrakanth, *Water Resource Economics*, DOI 10.1007/978-81-322-2479-2_1

boundaries. Ultimately, sustainability in groundwater extraction and use requires sacrifice in extraction in the short run and long run in line with the recharge.

1.2 How Much Water Does the Earth Receive?

The volume of water on the earth is 1400 million km^3 which can submerge the earth 3000 m deep. About 97.3 % of this water is salt water and only 2.7 % is fresh water useful for drinking/irrigation and other uses. Out of this 2.7 % freshwater, 75.2 % is in frozen form in Polar Regions, 2.2 % is available as surface water in lakes, rivers, atmosphere and moisture, and 22.6 % is available as groundwater[1]. Thus, groundwater resource is the nature's benediction and more so for hard rock areas (HRA) of India where the recharge is hardly 5 % to 10 % of rainfall. Groundwater is simply water filling spaces between rock grains or in cracks and crevices in rocks and these spaces are abysmally narrow. The rock layer that yields sufficient groundwater is called an aquifer. Aquifer may be a few feet or hundreds of feet thick; located just beneath the earth surface or hundreds of feet down; underlying a few acres or thousands of square miles. Groundwater does not occur downward all the way to the core of the earth. At some depth beneath the water bearing rocks, the rocks are water tight[2]. Obviously the volume of water held depends upon the ratio of open space to total volume (porosity). Many tend to think of groundwater as underground lakes or streams or as fossil water, which are extremely rare. The popular (mis) understandings with respect to groundwater even among the literate people is that (1) the resource flows continue without any effort to recharge (2) there is a pool of the resource from which one can pump as much as s/he wants to and (3) the flow increases with depth. These (mis)understandings, have resulted in the current predicament of groundwater (wells), where virtually most dug wells have failed or are failing gradually, drilling depth/s of wells are increasing over time, groundwater yield of wells falling over time, pump Horse Power used to extract groundwater is increasing over time, cumulative interference among wells is increasing due to violation of isolation distance and mushrooming of wells causing economic suffering for marginal and small farmers who cannot afford huge investments on drilling well/s. The National Aeronautical Space Agency NASA study highlights only the groundwater depletion in Punjab, Haryana and Rajasthan in the map below, discounting the predicament of groundwater depletion in the hard rock areas of India, especially the Deccan Plateau. The recent study from NASA (http://www.deccanherald.com/content/484290/groundwater-depletion-india-worst-world.html) highlights that "Groundwater is disappearing fast from the world and India is among the worst hit, shows data from NASA's Gravity Recovery and Climate Experiment (GRACE) satellites. Among the world's largest groundwater basins, the Indus Basin aquifer of India and Pakistan, which is a source of fresh water for millions of people, is the second-most overstressed with no natural replenishment to offset usage,

[1] Information System Directorate Performance Overview 1996.

[2] American Institute of Professional Geologists 1980.

said two new studies led by the University of California - Irvine (UCI), using data from GRACE satellites."

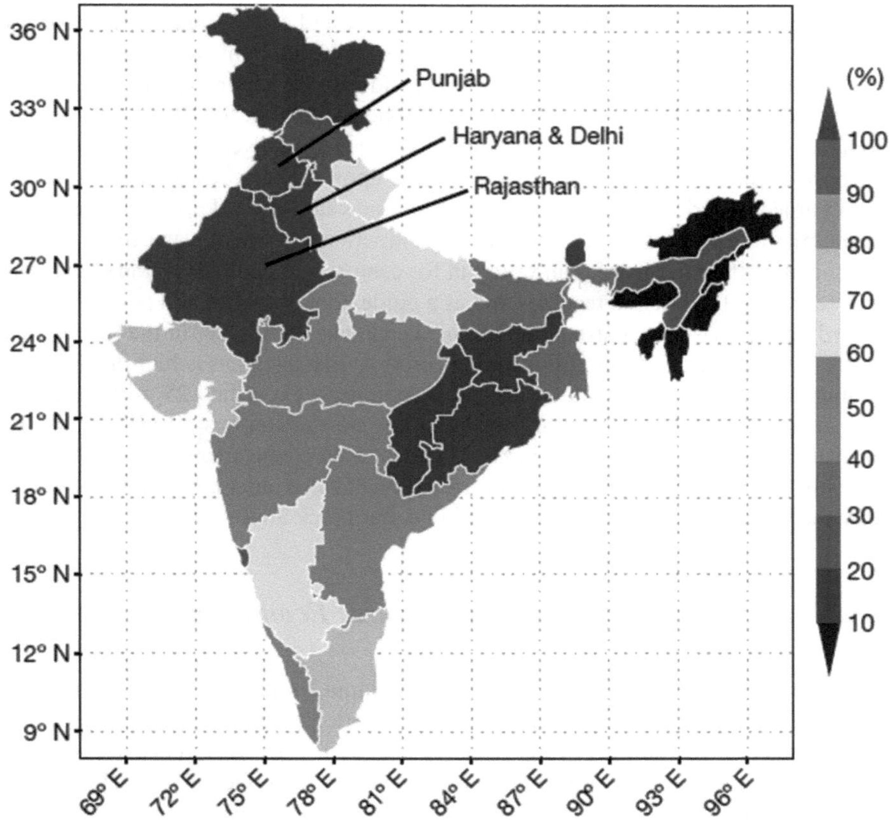

1.3 Lessons and the Purpose of This Book

Out of the earth's freshwater (of 2.7 %), even though the share of groundwater (22.6 %) is more than the surface water (2.2 %), the constraints to tap groundwater are far more compared with the surface water; viz., groundwater is not visible, is a function *inter alia* of the degree of weathering, type of rock (hard rock or alluvial..), type of aquifer (confined, unconfined, semi confined…), location (close or away from the farm), topography (flat, undulating, sloppy terrain…), depth of availability, degree of cumulative interference (low, high), availability of power to lift groundwater, effective demand for the use of groundwater (for irrigation or other uses), level of investment by user (farmers, households, others). While for tapping surface water public investment at macro level is required, while that for groundwater, private investment at micro level is crucial, since property rights to surface and groundwater are different. Whatever be the *de jure* (or legal) rights assigned to groundwater, *de facto* (in practice), private property rights continue to hold, since the transaction costs of administering rules is prohibitive for any government given the large number of scattered small holdings all over the country and ever increasing demand for groundwater for different uses by different users. Therefore, unless frantic efforts are made to educate the farmers and users of groundwater to understand and appreciate the hydrogeological, biophysical and socioeconomic factors for sustainable or reasonable use of groundwater, groundwater continues to be overdrafted in India. The purpose of this book is to educate and impress upon farmers, researchers, students, policy-makers and all users of water and groundwater, the role of water resource economics as a process towards sustainable extraction and use of water for maximizing social welfare not just for an year, but for centuries to come. It is hoped that this will serve as a textbook to teachers and as a guide book for users and policy-makers regarding the 'wise' or 'reasonable' use of scarce water resource. In the process the predicament with respect to groundwater resource management is highlighted. In a country with such a large number of actors (pumpers) exceeding 25 million spread over across the length and breadth of India, the transaction cost of any regulatory measure is colossal. Hence, the first requirement is to create awareness and educate not only the farmers pumping water for irrigation but also other consumers regarding how fragile is the groundwater resource in the hard rock areas of India.

1.3.1 How Water Is Treated in Resource Economics Perspective

Water for irrigation is treated as a 'factor' or 'input' in production in economics/agricultural economics. However, water for irrigation needs to be treated as a 'resource' in natural resource or water resource economics. The difference between the two can be understood by appreciating how treatment as a 'factor' is myopic, while treatment as a 'resource' is holistic. For instance, an officer can get a labour with a calling bell. Instead, if the officer calls the labour by his/her name, a specific person comes for help with all wherewithal love and affection. Hence while economics refers to 'labour' as input, in resource economics, labor is referred to as 'human resource'. Similarly, while economics refers to 'water for irrigation' as a factor of production, in resource economics, it is referred to as 'water resource'. It is in order to discern the differences between economics/agricultural economics, natural resource and environmental economics in their ten attributes (Table 1.1). Towards optimal use of a factor, while economics uses the rule MR = MC, resource economics uses MR = MC + OC. Where MR = Marginal return, MC = Marginal cost and OC refers to the opportunity cost, user cost, royalty, shadow price or resource rent. OC refers to the cost imposed on the future for having used a unit of the resource at present. Here there is an assumption that the resource under consideration is an exhaustible resource. Even though groundwater resource cannot be strictly considered as exhaustible resource, the way groundwater is getting over extracted in relation to natural recharge, there are no compelling reasons to disbelieve that groundwater can be treated as an exhaustible resource, more so in hard rock areas. In fact the

Table 1.1 Table key differences between economics, natural resource and environmental economics

Features	Economics	Water resource economics
1. Subject deals with	Goods and services produced by human beings and consumed by others including humans	Resources and services from nature consumed by human beings, non-human beings
2. Example	Pen, paper, cloth and labour	Water, fishery, forests, environment, minerals, environmental services, biodiversity services and pollination
3. Property rights	Property rights to goods and services are well-defined	Property rights to natural resources and natural resource services are ambiguous and are not properly defined/definable
4. Focus with examples	Focus is narrow, on inputs. Examples: irrigation, crop area, timber and fish catch	Focus is broad/wide on natural resource. Examples: water resource, land resource, soil resource, forest resource, fishery resource and biodiversity services
5. Limitation	As the focus is on input use, user is interested in efficient use of input and is not much concerned with scarcity, externality, equity and sustainable use aspects	As the focus is on natural resource extraction and use, user, in addition to efficiency, is also concerned with scarcity, externality, equity, sustainable use aspects, reuse, recycle and ethics of resource use
6. Good/commodity	Product focus is on output. Ex. paddy/jowar/sugarcane/timber. Only 'good' goods, like pesticide, for instance are considered. 'Bad' goods like pesticide pollution are not considered. Similarly ecological/watershed functions are not valued	Product focus is on throughput which includes 'good' good (pesticide) as well as 'bad' good (effluent); all goods (timber, non-timber products, as well as ecological and watershed functions
7. Production function	Production function has increasing and decreasing returns, reaching maximum output and diminishing returns. Thus, there is an irrational, a rational and an irrational stage. Decision rule is given by marginal return = marginal cost and where marginal return cuts marginal cost curve from above	Production function varies with each natural resource. For e.g. Stumpage value curve in forestry has no maximum output and decision regarding optimal rotation age depends on choice of interest rate. Decision rule is marginal return = marginal cost + opportunity cost (user cost). User cost is cost imposed on future for having used a unit of resource at present
8. Market	Market imperfection may exist	Along with market imperfection, market failure and missing market also exist

Table 1.1 (continued)

Features	Economics	Water resource economics
9. Price/value	As property rights and markets are well-defined, market price is relevant. The total economic value includes only use value	As property rights and markets are not well-defined, 'value' rather than 'market price' is relevant. Thus, total economic value includes use value, nonconsumptive use value and nonuse values
10. Costs/returns	Marginal private cost and marginal private benefit are relevant for decision making	In addition to private costs and benefits, social costs and benefits are also relevant for decision making

renewable resources such as groundwater, forests, are behaving like exhaustible resources, and exhaustible resources such as iron ore, fossil fuels are behaving like renewable resources since price of exhaustible resources is falling, while the price of renewable resources such as timber, water are rising. Hence, resource economics uses the present value of the cost imposed on the future and adds to the marginal cost (MC) of pumping groundwater for instance in order to compare with the MR from groundwater to obtain the economic optimal level of groundwater use. In the case of commodities, inputs, factors, as the market is relatively well-defined in terms of property rights, the market price exists. For example, fertilizer price, seed price or labour wage exists due to market for seed, fertilizer and labour. However, price of land and price of water are not as clear as the price of seed or fertilizer due to lack of well-defined property rights on resources. Hence, it is the value of a resource versus market price of input. The value of a resource is subjective and differs from farmer to farmer in, the case of groundwater as the resource cost depends upon the depth of drilling, number of initial and premature well failures. However, market price of seed is relatively uniform (Table 1.1).

1.4 Why Water Is Vital for India's Agriculture

Water is the lifeline of India serving as productive and protective input to feed the mammoth population of 1.2 billion, growing at 1.76% (2008). With 329 ml ha of geographical area, the net cropped area forms 43% and gross cropped area forms 59%, the highest in the world. Area under food grains account 63% of the gross cropped area. About 45% of the cropped area and 48% of food grain area are irrigated. Rice production forms 42% and wheat 35% of the total food grain production of which 59% of rice and 91% of wheat are from irrigated area contributing to food security, stability, productivity and resilience. Considering individual crops, 94% of sugarcane area, 91% of wheat, 75% of barley, 74% of rapeseed and mustard, 59% rice, 35% of cotton, 34% of gram, 31% of sunflower, 25% maize and 21% groundnut are irrigated. For India's green revolution, irrigation contributed 60 % to the growth of agricultural productivity. Around 60 % of human body weight is water. Appropriately India's green revolution can be termed as India's groundwater exploitation revolution.

1.5 Himalayas Are 'The Roof of the World'

Himalayas bordering India referred to as 'The Roof of the World', have the largest area of glaciers and permafrost away from the poles. Here, ten large rivers of Asia take birth. Because of these, the Indo-gangetic plains are better placed with regard to both surface water and groundwater due to alluvial soils receiving both rainfall and snowmelt water. The rest of India dominated by HRA , form 65 % of the geographical area, is starved of both groundwater and surface water for all needs. It is here that the demand for water surpasses the supply in leaps and bounds resulting in groundwater overdraft. Here to meet the water needs of the growing population, both agriculture and human health goes astray, unless concerted efforts are made towards conservation and wise use of water across uses and users, which is the very message of this book.

1.6 Monsoons Are Decisive for India's Economy

Though around 50 % of cropped area is irrigated, and 80 % of this irrigation is from groundwater, from a rainfall of 750–1500 mm, India's water resources are totally dependent on monsoon, since both groundwater and surface water are fed by monsoonal rainfall. Thus, the quantum and distribution of rainfall is a major determinant of the farm economy, irrespective of whether the farm is rainfed or irrigated. Thus, both surface and groundwater are dependent on the monsoons. More than 90 % of water is used for irrigation. Thus, rainfed, surface water irrigated and groundwater irrigated agriculture suffers from the vagaries of monsoon. The climate change exacerbates the predicament. Irrigation efficiency in general and economic efficiency in the use of irrigation water in particular, shapes the economy of the farming sector.

1.7 The HRA and 60 % Syndrome

About 65 % of India's geographical area constitutes the HRA fraught with low recharge ranging from 5 to 10 % of the rainfall. The groundwater irrigation in these areas constitutes about 60 % of irrigated area. Around 60 % of irrigated area here contributes to food production. Thus, groundwater irrigation holds the key to food security of India.

1.8 Public Investment in Surface Water and Private Investment in Groundwater

In the surface water irrigation provided by irrigation tanks, dams, reservoirs and canal systems, the entire investment is borne by the public, while farmer does not bear any cost of infrastructure and is expected to remit the modest water charges as water rate. As Water Users Cooperatives are unable to convince farmers to pay for the modest 'water rates' (equivalent to ₹ 100/acre for paddy and ₹ 400/acre of sugarcane and ₹ 66/acre of semidry crops in Karnataka) farmers often do not remit the water rate/water charges for surface water, virtually treating surface water as a 'free' public good. In addition, due to inefficiency in the use of surface water, water logged, saline and alkaline soils are increasing. In the case of groundwater, however, the farmer invests on groundwater well, irrigation pumpset and accessories. In addition, the groundwater well farmer bears the risk of initial/premature well failure, pump set, electrical fixtures, installation, conveyance pipes, and other accessories, including drip/sprinkler irrigation, if any. In addition, farmer is required to pay for the electricity to pump groundwater for irrigation. Farmers in most of the states in India do not pay for this electricity. Thus, both surface and groundwater irrigation sector is fraught with low revenue—high investment muddle and the associated negative externalities. Relatively groundwater farmers are efficient compared with the surface water farmers, since groundwater farmers contribute at least towards the bulk of the investment which goes towards irrigation well and pump.

1.9 What Is Well Interference and Cope of Depression and How it Affects Groundwater Extraction?

Aquifer is a water-bearing rock which transmits water to wells and springs.Rainfall (recharges or) adds water to porous rock of aquifer and this rate of recharge varies with aquifers, and is a crucial parameter to be considered while pumping water from a well. Obviously indiscriminate pumping draws down the water fast in the aquifer and causes a well to yield less and less water and eventually go dry due to 'cone of depression' (Fig. 1.1). Pumping in one well too much and too fast can dry up the neighbouring well located in a common aquifer due to well interference.

1.10 Consumptive Use

Water used in irrigation/agriculture is referred to as 'consumptive use'[3] as irrigation water once applied, cannot be recovered akin to domestic and municipal water. Water in uses other than irrigation, is 'nonconsumptive use' and can be recovered.

---

[3] Consumptive use is water applied to crops or livestock that evaporates and not returned to the immediate environment. All water used indoors can be recycled and hence called nonconsumptive

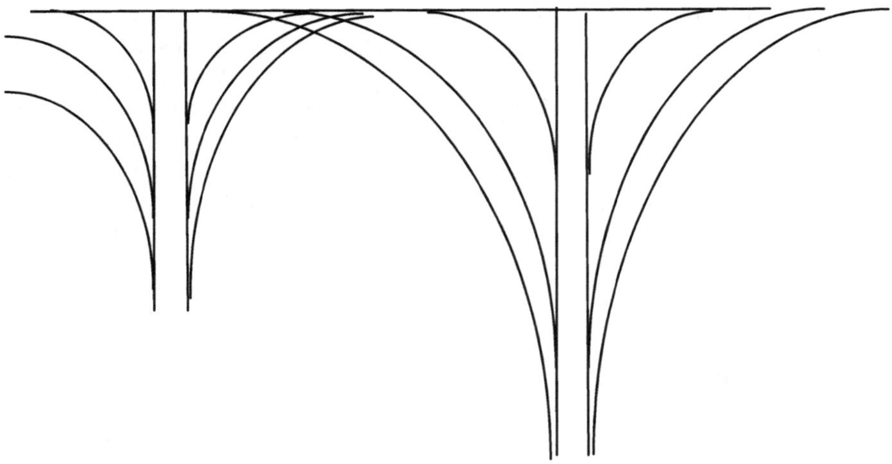

Fig. 1.1 Well interference and cone of depression

Both surface water and groundwater are complementary. Agriculture or irrigation being the world's largest use of water, water-use efficiency is thus relatively more important in agriculture than other sectors, since water savings in agriculture can make it amply available for other uses/users. Prima facie indicators of inefficiency in surface water use are relatively low output and low returns per unit volume of water and per acre of irrigated area, increase in water logged, salinity and alkalinity area. Prima facie indicators of groundwater inefficiency are *inter alia* raising costs of groundwater extraction, externalities, increased probabilities of initial and premature well failure, reduced life and age of irrigation wells, reduced groundwater yield of wells, and increased depth of drilling to groundwater.

1.11 India—Top Extractor of Groundwater in the World

India gets around 4000 km^3 of rainfall annually within 100 h of downpour, a major portion of which is in eastern India. Thus, water storage and its availability for agriculture are crucial for India. The number of irrigation pumpsets in India increased from 0.15 million in 1950s to around 19 million by 2000 and is annually pumping 220–230 billion m^3, twice that of the USA and six times that of Western Europe, ascending as the *world's largest extractor* of groundwater. There has been an exponential growth in groundwater extraction from 37% in 1998 to 58% in 2010 pumping an average 75 acre-inch of groundwater per well per annum with the number of wells rising from 0.1 million in 1960 to 25 million in 2010.

use. But water used outdoor (say agriculture) cannot be recycled due to evaporation, hence called 'consumptive use'.

According to International Water Management Institute (IWMI)[4], 'Currently, over 80 % of irrigated agriculture in India is supported by groundwater. India has 25 million pumps with almost a million added every year. With little control on groundwater extraction, the country faces serious groundwater depletion. Food grain production has increased as more land is brought under cultivation. High-yielding crop varieties are grown that also have high water requirements, which are being met through a boom in groundwater use. The physical expansion of large-scale irrigation schemes has reached its limits. Innovative water-saving techniques and improved water-use efficiency are essential'.

1.12 Carbon Emissions

With the advent of shallow and deep tube well technologies, dug wells where manual lifts were being used, are no longer common in India. Thus, groundwater has to be pumped from depths beyond 500 ft in peninsular India using external source of (largely thermal) energy. With 25 million irrigation wells (pumpsets) in operation in India, the use of electricity and diesel is responsible for 16–25 million t of carbon emissions forming 4–6 % of India's total carbon emission. Thus, the groundwater hotspots are western and peninsular India, which are crucial for both climate change mitigation and adaptation[5].

1.13 Physical Properties of Surface Water of Relevance to Economics

Water is a natural resource from rainfall and snowmelts. In India, it is largely through rainfall, which varies considerably across different agroclimatic and ecological zones, across seasons. Surface water accumulates in ponds, tanks, dams, reservoirs flowing through rivulets, and rivers before joining the ocean. Surface water is a 'visible resource' and hence property rights are relatively well-defined, as they transcend across states. Thus as some rivers originate in one state, flow through other states and join the ocean in some other state, riparian laws come into force. The final award of the Cauvery water disputes tribunal[6] among states of Karnataka (where the river takes birth), flows through Kerala and reaches to ocean in Tamil Nadu, is again questioned in the Supreme Court, as Karnataka state opines that the tribunal has not addressed the water allocation during distress and groundwater potential has not been considered. Therefore enforcing property rights has a transaction cost for the state/s.

---

4 http://www.iwmi.cgiar.org/Publications/Water_Issue_Briefs/PDF/Water_Issue_Brief_%2011.pdf.

5 Shah 2009, *op cit.*

6 http://wrmin.nic.in/index3.asp?sslid=393&subsublinkid=376&langid=1.

1.14 Physical Properties of Groundwater of Relevance to Economics

Groundwater, a natural resource, which is also fed by rainfall, poses challenging tasks for governance due to ambiguous property rights. Currently, as most 'dug wells' have vanished, groundwater is being tapped by borewells, where it is an 'invisible' resource. In HRA, where cumulative interference among wells due to violation of 'isolation distance'[7] is a rule than an exception, the property rights are intricate. Even though the user (farmer/consumer) makes a private investment to tap the groundwater, there is no guarantee that she/he strikes water and will continue extracting the water for a fairly longtime period. Thus, the life and age of the wells *inter alia* depend upon the probability of initial and premature well failure which are influenced by cumulative interference, cropping pattern, depth and rate of extraction and efforts to recharge.

1.15 Water as an Economic Good: Role of Literacy and Governance

In developed countries where literacy and governance are appreciable, water is possible to be treated as an economic good, for the reason that water be used in a 'wise' rather than a 'beneficial' manner. In situations where literacy and governance are poor, and where political economy considerations overrule the basic nature of the natural resource, it is challenging to make users understand the scarcity value of water. It is not totally possible to treat water as an economic good, since water is an essential resource which is required by humans, livestock, wild animals and all the flora and fauna. The danger of treating water as an economic good is that it becomes unethical to price or value water for non-humans. However, farmers should be trained and educated in such a way that they make 'wise' use of water rather than 'beneficial' use of water. How to understand this? Suppose a farmer has a choice of cultivating paddy/sugarcane or banana given the market, it is wise to cultivate banana under drip irrigation, since banana uses less water than paddy/sugarcane and also fetches higher net profit. Similarly, if there is a choice of cultivating vegetables or flowers, it may be wise to cultivate flowers, as flowers use less water and yield great net revenue.

1.16 Sustainability in Water Use Requires Sacrifice

Sustainability in groundwater extraction and use requires sacrifice in extraction in the short run and long run in line with the recharge. In common parlance, efficiency refers to achieving the level of input use, where the marginal returns from the use

[7] Usually the interwell distance between two borewells is around 850 ft as recommended by the National Geophysical Research Institute NGRI, Hyderabad http://www.ngri.org.in.

of groundwater matches its marginal cost. However, this criterion does not consider equity, since it considers efficiency of each farmer independent of the neighbouring farmer/s. Efficiency also ignores sustainability, since it considers the present extraction without concern to needs of the future. Unlike other areas of the world, farmers in HRA of India, need to extract and use groundwater in a sustainable manner. This requires sustainability in groundwater extraction and use in the framework of natural resource economics.

1.17 What Is Sustainable Extraction of Groundwater

Following the principle of marginal cost pricing, which determines the optimal input use, at zero price of groundwater extraction (Fig. 1.2), OZ is extracted since groundwater is virtually free. However, if the farmers are made to pay for the extraction cost, OA* is extracted. But the price includes only the cost of extraction, ignoring other costs. Thus, if the farmer, in addition, is charged the user cost or royalty, which is equal to the return from groundwater foregone in future for having extracted an unit volume of groundwater at present, OB* is extracted. This level can be considered as natural resource economic optimal since it includes the marginal cost of extraction as well as the user cost (the dynamic component). Extraction further adds environmental costs in terms of poor groundwater quality, upon addition of environmental costs, the extraction level reduces to OC*. Addition of ecological costs through secular overdraft of groundwater and associated effects results in further reduction of groundwater extraction to OD*. And lastly the addition of sustainability costs of groundwater extraction, which requires sacrifice on the part of farmers in groundwater extraction considering not only his/her needs but also those of neighbour's and societal needs, not only at present but also in the future,

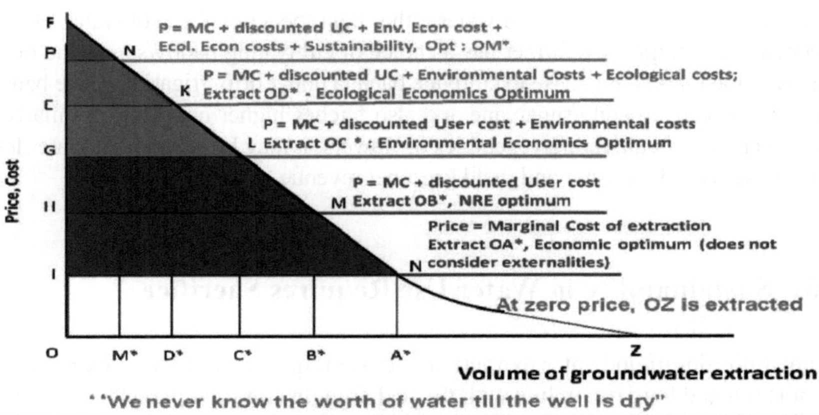

Fig. 1.2 Water resource economics—heuristic net social benefits from groundwater use according to economics, natural resource economics, environmental economics and ecological economics

incorporating, economic, environmental and ecological costs, will further reduce the extraction to OM*. The role of technology, institutions and markets in achieving sustainable use of groundwater is crucial.

1.18 Property Rights to Groundwater with the Advent of Borewells

Occurrence of groundwater in the unconfined aquifers, in the HRAs, is highly sensitive to interactive effects of wells and renders groundwater as a Wantrupian fugitive resource. Such aquifers yield water by draining of materials near the well. In the HRAs about 90 % of the aquifers are unconfined. The nature of groundwater rights is intricate and the rights are dynamic and are functions of the demand and supply side forces determining the availability of groundwater.

The property rights to groundwater are dynamic and change with supply and demand side factors which jointly determine the property rights. Dynamic implies that the property rights change over time with the advent of new borewell/s. A farmer who is an early comer in groundwater irrigation, growing low water intensive crops lifting water from dug well with manual lifts, almost feeling that he is enjoying (permanent?) private property rights to groundwater, will suddenly be shattered once there emerge(s) neighbouring farmer(s) who tap groundwater from deeper layer(s) from borewell due to cumulative well interference resulting in permanent failure of the dug well. Such dug wells which were extracting groundwater sustainably in India, have virtually vanished due to the advent of borewells decoupled with recharge structures at least in the HRAs. Dug wells made groundwater visible to enable the farmer to plan what crop/s s/he could cultivate with the available volume of water, what is the extent of recharge every day and also across seasons, what type and capacity of pump need to be used. The advent of borewells made groundwater as well as extent of daily recharge invisible and hence robbed the information regarding proper crop plan and choice of the right type and capacity of pump to lift the groundwater. More importantly due to mushrooming of irrigation wells, the life and age of borewells drastically reduced when compared with that of dug wells.

1.18.1 How Definite Are Property Rights to Groundwater?

Groundwater is an indispensable resource for irrigation in many pockets of HRAs especially where there are no flows of perennial rivers. In India, the rights in groundwater belong to the land owner as groundwater is attached to the land. There is theoretically no limitation on the volume of groundwater extraction by a land owner. Since, landownership is prerequisite to ownership of groundwater[8], it is difficult

[8] Singh 1992.

to assign 'open access' nature to groundwater resource. Though land owners own groundwater *de jure*, this right is limited by the huge investment necessary to tap the groundwater by construction/drilling of irrigation well(s) and high well failure probability, which makes a selected few among them to have access to groundwater. Unless groundwater is tapped in a well and is available, there is no accessibility, since there is no guarantee that any land owner who attempts to construct/drill a well is assured of groundwater, even for a short period. Initial and premature failure and falling life of irrigation wells are common features in HRAs. For instance, in the eastern dry zone of Karnataka, the (Negative binomial) probability of well failure is estimated to be 40 %[9] in the 1990s, increased to 0.7 in 2013, which implies that a farmer has to drill at least two wells, one of which (may be) is successful that too for an unknown period of time. A common understanding is that when well exists, water should. However, it is not the case since even in the fourteenth century dried wells are recorded by St Purandara dasa![10]

Under these circumstances, the groundwater rights are obscure since groundwater is being extracted myopically not recognizing the fact that each one's extraction is not independent, but an interdependent function of the extraction in neighbouring well(s) at a time and overtime. This is leading to cumulative interference of wells which is reducing age and life of wells, the gross area irrigated by wells, negative externalities, transaction costs and aftermath. In HRAs, the occurrence of groundwater is highly sensitive to interactive effects of wells and renders groundwater as a fugitive resource. Ciriacy-Wantrup[11] opines that 'definite property rights (to groundwater) belong only to those who are in possession—that is who gets there fastest with the mostest'. Thus, the sustainability of groundwater institutions depends *inter alia* upon the nature of aquifer, volume of groundwater recharge and discharge, isolation distance between wells, recharge efforts, rainfall, age of irrigation well, crop pattern, number of pumpers, all influencing stock and flow of groundwater.

1.19 Open Access Resource?[12]

In India, the rights in groundwater belong to the land owner as groundwater is attached to the land property (Indian Easements Act 1882)[13]. Hence, strictly landowners have access to groundwater[14]. As only those who own land have access to

[9] Nagaraj et al. 1994.

[10] Bayarithu endu bhavi neerige pode, Bhavi jala batthi baridaitho hariye (I went in search of a dug well to quench my thurst, but I found that the well had dried), thus said, Purandara Dasa, the father of Karnatak music way back in the 14th century itself.

[11] Ciriacy-Wantrup (1968).

[12] Chandrakanth and Arun 1997.

[13] http://www.commonlii.org/in/legis/num_act/iea1882158/.

[14] Singh 1992.

groundwater, groundwater cannot be an open access resource. However as 92 % of the farmers own land, groundwater may become open access, if most or all of them have access to capital to drill wells. Thus, even though land owners own groundwater *de jure*, this private property right is limited by the volume of investment necessary in drilling irrigation well(s) which is a function of well failure probability, which makes *a few among them* to have both physical and financial access to groundwater *for a certain time period as the access is conditioned by cumulative interference among wells*. Even considering the fact that the price of electricity to pump groundwater is fully subsidized (as it exists/existed in India), unless the farmer has *physical access* to groundwater, even if electricity is available, s/he has no access since, initial and premature failures are imminent due to the sensitivity of cumulative interference in HRAs. In the eastern dry zone of Karnataka, in peninsular India, the probability of well failure is estimated to be 40 %,[15] which implies that a farmer has to drill at least two wells, one of which may be successful. Thus, can wells exist without groundwater? Hence, it is difficult to admit that 'groundwater is an open access resource' at the one extreme or that 'groundwater has private property rights' at the other.

1.20 Common Property Resource and Reciprocal Externalities

In HRAs, clearly definable boundaries are difficult to establish for groundwater unlike alluvial aquifers. Hence, groundwater poses challenges to be considered as common pool resource, and hence common property regime is not easily applicable. Further, in the HRAs, drilling irrigation wells is increasing at a compound growth rate of 10 %. The *density of wells per unit area as well as the number of wells per million cubic meter of groundwater*, which determine the degree of interactive effects of wells, are increasing over time. Under these circumstances the groundwater rights are obscure, since farmers with opportunistic behaviour are not realizing the fact that each one's extraction is interdependent on the neighbouring wells' extraction at a time and overtime. This results in reciprocal externality, since all the users of groundwater impose external costs on all other users including themselves. In the case of unidirectional externality, a farmer by drilling deeper and/or extracting higher groundwater, inflicts externality on all others. But this is difficult and expensive to be established. However, over time, all farmers pumping groundwater impose external costs on all others (including themselves) due to cumulative interference, which can be termed as reciprocal externality[16,17]. Under these

---

[15] Nagaraj et al. 1994.

[16] Dasgupta 1982.

[17] Results have provided ample evidence of the substantial interactive effects of wells and with the inclusion of negative externality costs, the annual amortized cost of irrigation works out to around ₹ 14,000 per farm.

circumstances it is difficult to assign common property rights *de jure* or *de facto* to groundwater[18]. Hence, in HRAs property rights to groundwater are nebulous. With such an ambiguous rights regime, groundwater literacy and capacity building are also the prerequisites for collective action towards wise use to achieve sustainable groundwater management at low transaction cost.

1.21 'The Six Blind Men and an Elephant'

Groundwater is the 'elephant' being explained by the 'six blind men' who are the (1) hydrogeologist/water diviner, (2) government (regulator/policy-maker), (3) power (utility), (4) researcher, (5) user (farmers/households) and (6) driller. The prima-facie evidence for such an observation is for groundwater, both recharge and extraction are estimates. For recharge, the Groundwater Estimation Committee (1984), revised in 1997 and the latest in 2009[19]. Perhaps many more revisions may still be desirable. But for groundwater extraction, since there neither are water meters nor electrical meters fixed on irrigation wells, the volume of groundwater pumped is just an estimation subject to umpteen queries. In between the estimations of recharge and extraction, the six blind operators of the groundwater elephant act with asymmetric information, transaction costs, negative externalities, political economy, weak governance and competing demands for the scarce resource. The interrelationships among factors impinging on groundwater recharge (supply side factors) and extraction (demand side factors) are thus a rigmarole. As groundwater and surface water are closely linked, planning and management of groundwater and surface water should go together as in river–basin approach focusing on conjunctive use as highlighted in Integrated Water Resources Management (IWRD).

1.22 Markets, Technologies, Institutions, Social Capital, Transaction Costs and Governance Structures

Water-use efficiency involves use of market (M), institutions (I) and technologies (T). Experience in India shows that in addition to M, I and T, the quality of social capital (SC), transaction costs (TC) and governance (G) are important. *A society that has farmers/consumers and policy-makers who can appreciate the hydrogeological and biophysical properties of groundwater and surface water, their quality of SC will reduce the TC of G to achieve wise use of the scarce water resource source using the right Is, Ts and Ms.* As water is indispensable, sustainable use is crucial and vital, and as water is the most vulnerable for climate change, the roles of M, I, T, G, SC and TC are crucial for sustainable water use.

[18] Marcus 1995.

[19] Ministry of Water Resources 2009.

1.23 Groundwater and Climate Change

It has been predicted that climate change will increase the demand for groundwater for agriculture and other uses as the droughts and drought periods would increase. Further, studies have demonstrated the scope for reducing carbon footprint of groundwater. For every meter decline in pumping water levels, greenhouse gas (GHG) emissions increase by 6%. For 1% increase in groundwater irrigated area, there is a 2.2% increase in GHG. Due to growth of 10% in groundwater irrigated area per year, the GHG emission is increasing at 22%. For 1% increase in the share of diesel pumps to total pumps, the GHG emissions reduce by 0.3%. For 1% increase in irrigation efficiency, the GHG emissions reduce by 2.1%[20]. Thus, in the context of climate change, drip irrigation and shift to low water—high value crops are crucial due to win–win situation as it saves groundwater use, releases less CO_2 while pumping groundwater and augments area irrigated. However, diesel and biodiesel are no proper substitutes for electrical power to lift groundwater from deep borewells. This is the reason, why diesel irrigation pumpsets are popular in northern India where groundwater is relatively available compared with southern peninsular India. However, increasing reliance on groundwater is disadvantageous since hard rock aquifers which form 65% of India's area have slow recharge, where pumping is energy intensive as water is extracted from deeper depths and increases carbon foot print.

1.24 Does Jevons Paradox Apply to Groundwater Extraction?

According to Jevons Paradox, technological progress which increases resource use, in fact tends to increase the rate of consumption of the resource rather than conservation. Whether groundwater will still be conserved remains a researchable question, till an answer is found by conducting surveys in different parts of India. However, drawing analogy from the situation in San Joaquin valley of California, USA, where even with acute scarcity of groundwater hardly 15–20% of farmers have shifted to drip irrigation, in HRA, the shift to micro irrigation will be slow and steady to the extent that full efficiency in groundwater use will be achieved. Even with provision of full incentives, it is unlikely that farmers will shift to micro irrigation, since the concept of micro irrigation for narrow spaced crops is yet to be developed. In pockets of Kolar district of Eastren dry zone of Karnataka, the use of drip irrigation has increased, since farmers are gradually realizing the economic value of groundwater. In addition, the size of the holding of farmers with irrigation wells is not substantially large for farmers to expand the area under drip irrigation, with the saved water. Thus, Jevons Paradox eschews groundwater irrigation in HRA of India.

[20] Shah (2009), op.cit.

1.25 Supply Side Factors Governing Groundwater

About 90 % of groundwater is used for irrigation and is extracted through (private) irrigation well/s. The factors which shape availability (supply) of groundwater are the number of rainy days, volume of rainfall, the nature of aquifer (confined or unconfined), the proximity to recharge points, the presence of dykes or groundwater barriers and lineaments, the surface water bodies, commitment to recharge efforts in addition to natural recharge, quality and horsepower (HP) of irrigation pumpsets, supply of electricity at regular voltage and others.

1.26 Demand Side Factors Governing Groundwater

The main factors which influence demand for groundwater are cropping pattern (cultivation of low water intensive food/cash crops versus high water intensive food/cash crops), demand for horticultural produce like fruits and vegetables, proximity of metropolitan centres, irrigation methods (flood irrigation/flow irrigation/ furrow irrigation/drip/sprinkler/micro irrigation systems), subsidized electrical power to lift groundwater; presence of a few number of well owners versus large number of well owners, early comer/late comer in groundwater irrigation, age of irrigation well, degree of well interference, type of well (dug well, borewell), type of groundwater extraction devices (manual lifts/power lifts with varying HPs of pumpsets), well density, number of wells per unit of groundwater and proximity to peri-urban area.

1.27 Wide Estimates of Water Extraction and Use of Electricity

There are conflicting estimates of use of electricity for irrigation in India. As there are no electrical meters installed to measure the electricity used by irrigation pumpsets, the use of electricity for irrigation, the groundwater pumped from wells (i.e., groundwater extraction) and groundwater recharge (assumed to be 5–15 % of rainfall) are all estimations. And hence, precision in such estimates are still a long way. For instance, the recent publication by the National Sample Survey Organization indicates that considering the farmer households using non-human energy for irrigation, 66 % of farmers used diesel irrigation pumpsets while 33 % used electric irrigation pumpsets[21].

Researchers working on energy-growth linkage need to be careful while using the electricity use data available from published sources in India. As indicated ear-

[21] National Sample Survey Organization 2005.

lier, this data only includes estimates (but not actuals) of electricity use for pumping groundwater for irrigation as no metered data exists. In the absence of such data, conclusions drawn on electricity use in groundwater, are subjective as also on energy—water nexus.

Similarly, the statistics regarding the proportion of land irrigated by groundwater and surface water also varies widely across studies. The latest study by IWMI indicates that more than 80 % of land irrigated is from groundwater. Study by Stephane de la Rue du et al[22] indicates that 60 % of the land is irrigated by groundwater. Obviously such percentages cannot increase by huge proportions in just a year.

Similarly, the energy used per irrigation pumpset (IP) is estimated to be around 5904 kWh/year for electrical IP set and 6638 kWh for diesel pumpset (Stephane de la Rue du et al.) while Tushaar Shah (2009), estimates this to be 5863 kWh/ha of irrigation per year (in Andhra Pradesh). Here one also needs to assume that an IP set irrigates around 1 ha of land in a year. There is thus no uniformity in reporting the electricity use in India for agriculture.

A detailed study by Jha and Sinha[23] highlights the art and science of groundwater in all types of aquifers of India in a comprehensive manner and the reader can find maps of different types of aquifers and the complexity of groundwater resource availability across aquifers.

1.28 Power Supply to Irrigation Pumpsets Is Treated as a Residual

Electricity use for pumping groundwater is often highlighted as the single cause for over-exploitation. Field work-based studies in the Department of Agricultural Economics sponsored by the Ford Foundation and the World Bank from 1994 to 2005 (for details http://www.toenre.com/) indicate that the reciprocal externalities faced by the farmers due to cumulative interference of irrigation wells unaccompanied by recharge efforts, are largely responsible for well failure in the HRAs[24]. As electricity meters are not installed for irrigation pumpsets, the power utility treats the power supply for agriculture/irrigation pumpsets as a residual. Hence, the total power generated less metered supply, less Transmission (T), less Distribution (D) losses is attributed to agriculture. In many cases even a part of T and D losses enter the residual attributed to agriculture power use. Thus, the quantum of power supply for irrigation pumpsets is estimated to be 35–40 % in most of the states in HRA and is not an objective proportion of the power use in agriculture.

[22] http://www.escholarship.org/uc/item/0f05n9cr.

[23] http://www.cgwb.gov.in/documents/papers/incidpapers/Paper%201-B.M.Jha.pdf (perhaps 2008).

[24] Chandrakanth 2002.

1.29 Who Bears the Brunt of the Cost of Groundwater Despite Power Subsidy?

Considering the investment on irrigation borewell and pumpset on conservative basis to be around ₹200,000 (3000 €) (in 2013) and considering the proportion of well failure which is around 40 %, suppose the borewell serves for around 5 years at zero interest rate, the amortized cost of irrigation works to ₹ 40,000 (=3000/5=600 €)/ year. A typical borewell pumps around 1500 gallons of water per hour. Usually on an average, the electricity to pump groundwater is available for 4 hours/day. Considering the number of rainy days in a year to be 65 days, the farmers will put on the pump for say 300 days, extracting (300 days* 4 h/day*1500 gallons/h=) 1,800,000 gallons or 80 acre-inch of water per year. Even if this is taken as 100 acre-inches (or 100 ha cms), the irrigation cost, without considering the cost of pumping, thus works to ₹ 40000/100 acre-inch or ₹ 400 (5.8 €)/acre-inch. It is estimated that 42 kWh of electricity are required to pump 1 acre-inch of groundwater[25] from shallow borewell. However, with the increase in the proportion of deep borewells, the cost of electricity goes up and needs to be estimated for each area/each well.

On an average, around 6000 kWh are used per well to pump irrigation water annually which amounts to around 60 kWh/acre-inch of water pumped[26]. The average cost of electricity generation is around ₹ 3.68/kWh. The total subsidy works to around ₹ 22,000/irrigation well. Utilizing the data of 42 kWh/acre-inch for 100 acre-inch at ₹ 3.68/kWh (equivalent to the cost of generation of power in Karnataka during 2012), the cost of electricity works to ₹ 15,456. Thus, the electricity cost is around ₹ 15,000–22,000/year. However, the farmer incurs ₹ 40,000 and an addition of electricity cost pushes up the cost of pumping water to ₹ 60,000 (considering a sum of ₹ 20,000/irrigation pumpset, which forms around 33 % of the total cost. As a welfare measure, power for irrigation cannot be treated on par with other sectors. Obviously the power for irrigation is charged substantially lower than that charged for other sectors. For instance, given the cost of generation of power at ₹ 3.68/kWh, the Karnataka Power Transmission Corporation Limited (KPTCL) may value the power for irrigation at ₹ 1/kWh following the recommendation of the National Council of Power Utilities. If the power tariff is around ₹ 1/kWh, then the power subsidy forms around 13 % of the total cost of groundwater. Thus, farmer bears the major portion of the cost of water as she/he bears between 67–87 % of the cost of groundwater (depending upon the power tariff considered). The total cost of groundwater including electricity charges to lift groundwater varies from ₹ 460 to 600/acre-inch or Rs. 600 per ha cm. (2013) depending on power tariff. Thus, groundwater resource is crucial and valuable resource for farmers and rather than focusing on subsidy on electricity, it is desirable to focus on capacity building of farmers creating awareness with regard to

[25] Chandrakanth et al. 2001.

[26] Lecture by Dr WM Shivakumar, Executive Engineer and Research Officer, KPTCL to Ford Foundation sponsored project, Dept of Agricultural Economics, UAS Bangalore, 12 July 2004.

wise use of groundwater. Reducing electricity subsidy does not necessarily imply that groundwater extraction gets reduced, since in economic sense, the marginal productivity of groundwater and its equivalence to the ratio of the cost of groundwater to the price of the output determines the optimal use of groundwater in a myopic sense. More than 66% of IP sets in India do not use electricity, but, use diesel to extract groundwater (NSSO 2005).

1.30 Water Organizations in India

Water, land and forests are the three natural resources which have the maximum number of institutions and organizations governing their use (Table 1.2). Yet they are unsustainably managed and used. This shows the needs for proper governance of natural resources despite the existence of institutional structure. Considering only water, there are 16 organizations responsible for its wise use, still the 'wise use' has been a dream for end users as well as policy makers. It is a pity that groundwater as an input or as a resource is not accounted or costed (priced) in the computations of the Cost of Cultivation scheme (of the Government of India), as the RT (record type) forms do not include any information pertaining to drilling depth, casing, year of initial failure or premature failure (if any), number of previously drilled wells and their investment particulars and so on. Thus, even in the MSP (Minimum Support Price) announced by the CACP (Commission for agricultural costs and prices) the price of output offered to procure which is based on the cost of cultivation, does not include the cost incurred on groundwater resource by farmers. To that extent, the cost of cultivation of crops based on groundwater use, is underestimated, thus artificially over estimating the net returns! The cost of groundwater irrigation formed around 15% to 30% of the cost of cultivation of crops (on sample farms) in HRAs. Accordingly the net returns are over estimated by 15% to 30%.

1.30.1 Estimation of Reciprocal Negative Externalities

The water-use efficiency is obtained by first estimating the negative externality in well irrigation. The reciprocal externality or the negative externality is the difference between the amortized cost of irrigation per functioning well minus amortized cost of irrigation per well (Diwakara and Chandrakanth 2007). The hypothesis is that if all wells on the farm are functioning, then there is no externality and hence the amortized cost per functioning well will be the same as amortized cost per irrigation well (which includes both functioning and non-functioning wells). If the farm has both functioning and non-functioning wells, then, depending on their respective proportions of investments lost, the negative externality varies.

Table 1.2 Organizations monitoring water resource in India. (Source: World Bank 2010)

Level	Organization	Main functions
Central water development and management	1. Central ground water authority	Established in 1997, following supreme court orders, mainly to regulate, control, manage and develop groundwater resources in the country and supports states
	2. Central ground water board	Established in 1950 for dedicated groundwater research and monitoring, to support overall planning for development of groundwater resources and to provide support to states
	3. Ministry of energy and power	Clearance of hydro power projects
	4. Ministry of commerce and industries	Policy decisions and water use norm setting on water related to industry
	5. Ministry of environment and forests	Environmental clearance from MOEF for water related projects from environment (protection) act 1986
	6. Ministry of rural development	Rural development, land resources and drinking water supply
	7. Tribunals	Approval/sanction of water related projects for interstate rivers
	8. National water board	Established in 1990 under ministry of water resources, apex organization with responsibility for progress achieved in implementation of national water policy and other issues, reports to national water resources council
	9. National water resources council	Established in 1983 with prime minister as chair, minister of water resources as vice-chair, and concerned Union ministers/ministers of state, chief ministers of all states and lieutenant governors of union territories with secretary of Ministry of Water Resources as member secretary
	10. Ministry of water resources	Setting policy guidelines and programs for development and regulation of the country's water resources, but functions specific to groundwater resources through central ground water board
	11. Oil and natural gas commission	Member of central ground water authority and supplements deep well logging information
Central financing institutions	1. Rural electrification corporation	Development financing institution that finances, fully coordinates and oversees special project agriculture
	2. National bank for agriculture and rural development	Responsible for refinancing and standardizing substantial part of private sector groundwater

Table 1.2 (continued)

Level	Organization	Main functions
State	1. State electricity boards	Single window to individual farmers for obtaining pump set energization
	2. State government departments	Principally responsible for water use and control, as water is primarily a state subject
Local	Panchayats	Rural water supply, but to be devolved more water services and water resource management functions

1.30.2 Water-Use Efficiency

Water-use efficiency is measured in terms of crop output and net revenue realized per unit volume of water used by farmers. Prima facie can be said that those farmers who use drip irrigation, the water-use efficiency will be higher as they realize higher output per unit volume of water compared with conventional irrigation (or flow irrigation) farmers. This further has ramifications on increasing the resilience of farmers in the wake of well failure and thereby fight poverty and unemployment through increased income earning opportunities and the resulting social capital formation. The water-use efficiency can also be computed as the net return per acre of land. In the case of drip irrigation, the water applied will be low and hence the output per unit volume of water will be high.

The incentives provided for groundwater irrigation such as for credit, electricity, drip/sprinkler/micro irrigation, diesel, operation of schemes for groundwater extraction such as 'million wells scheme', ganga-kalyan yojana, macro schemes such as watershed development program, National Horticultural Mission, Tank rehabilitation program (Jala Samvriddhi Yojana Sangha JSYS), capacity building programs of the Departments of Agriculture, Horticulture, Krishi Vignana Kendras KVKs and State Agricultural Universities (SAUs) influence the water-use efficiency. In this book, a modest attempt is made to focus on the conclusions obtained from the research results obtained in groundwater and surface water resource economics in India.

References

American Institute of Professional Geologists (1980) Groundwater, water resources library. University of California, Berkeley

Chandrakanth MG (2002) Valuation of natural and environmental resources: methodology and estimation, Rapporteaur's report. Indian J Agric Econ 57(1):23–30

Chandrakanth MG, Arun V (1997) Externalities in groundwater irrigation in hard rock areas. Indian J Agric Econ 52(4):761–771

Chandrakanth MG, Shivakumaraswamy B, Sathisha KM, Basavaraj G, Adya Sushma, Shyamasundar MS, Ananda KK (2001) Paying capacity of farmers considering cost of groundwater and electricity in Karnataka. Paper presented to Karnataka Electricity Regulation Commission, 20th and 21st Aug, Bangalore.

Ciriacy-Wantrup SV (1968) Resource conservation: economics and policies. Division of agricultural economics. University of California, Berkeley

Dasgupta P (1982) The control of resources. Harvard University Press, Cambridge

Diwakara H, Chandrakanth MG (2007) Beating negative externality through groundwater recharge in India: a resource economic analysis. Environ Dev Econ 12:271–296 (Cambridge University Press)

Groundwater Estimation Committee (1984) Report of the Groundwater Resource Estimation Committee—groundwater resource estimation methodology. Ministry of Water Resources. Government of India, New Delhi

Information System Directorate Performance Overview (1996) Water and related statistics, central water commission. Government of India, New Delhi, p 1

Jha BM, Sinha SK (2008) Towards better management of ground water resources in India. Central ground water board, New Delhi

Marcus Moench (Ed) (1995) Groundwater management: the supply dominated focus of traditional, NGO and Government efforts, VIKSAT, Thaltej Tekra, Ahmadabad. Several studies presented at the workshop on "Water management: India's groundwater challenge" at VIKSAT, Ahmadabad in December 1993 have attributed 'common property rights' to groundwater resource.

Ministry of Water Resources (2009) Ground water resource estimation methodology, New Delhi. http://cgwb.gov.in/documents/gec97.pdf

National sample survey organization (2005) Situation Assessment Survey of Farmers: some aspects of farming, NSS 59th Round (January–December 2003), Report No. 496(59/33/3) National Sample Survey Organisation, Ministry of Statistics and Programme Implementation, Government of India

Nagaraj N, Chandrakanth MG, Gurumurthy (1994) Borewell failure in drought-prone areas of Southern India: a case study, Indian J Agric Econ 49(1):101–106

Shah T (2009) Climate change and groundwater: India's opportunities for mitigation and adaptation. Environ Res Lett 4(3):1–13. 035005 http://iopscience.iop.org/1748-9326/4/3/035005/pdf/erl9_3_035005.pdf

Singh C (Ed) (1992) Water law in India. The Indian Law Institute, New Delhi

World Bank (2010) Deep wells and prudence: towards pragmatic action for addressing groundwater overexploitation in India, Report 51676

Chapter 2
Externality in Irrigation

2.1 Externality

Externality is an elusive concept, as the experience does have unique effects on different individuals facing externality depending upon each one's perception. Some individuals do express externality, while some do not and the threshold levels of feeling external effects differ across individuals. If there are two individuals (A and B), the conditions for presence of externality are, (1) that action of a person (B) should enter the production/consumption function of a person (A) and this action should be a side effect on A (not main effect); (2) that action of a person (B) should be unintended or is not deliberate; (3) that there should be no contract or institution between the persons A and B governing the impact (meaning that if person B's actions harp on A, there is no rule that A can hurt B as tit for tat or in return); (4) that there is no price system between individuals A and B governing the impact (i.e. B does not compensate A for his/her actions and vice versa). Positive externality results in inefficiency (due to under production) and welfare loss, and negative externality results in inefficiency (due to over production or over extraction) and welfare loss. In the case of actions which are deliberate, then these are dealt through criminal law. Thus, for externality, the harm inflicted needs to be resolved through regulation, civil law or private negotiation. In this chapter, externality in groundwater irrigation is discussed in detail with a view to estimate them.

2.1.1 Reciprocal Externality

In hard rock areas (HRAs) each one's extraction of groundwater is not independent, but is interdependent on the extraction by neighbouring well(s) at a time and over time. This results in reciprocal externality, as all the users of groundwater impose external costs on all other users simultaneously and over time. In the case of unidirectional externality, a farmer by drilling deeper and/or increasingly extracting groundwater, inflicts externality on others and on himself at a time and over time due to interference of well/s. Over time, all farmers pumping groundwater, impose

M.G. Chandrakanth, *Water Resource Economics*, DOI 10.1007/978-81-322-2479-2_2

external costs on all others, including upon themselves due to cumulative interference, which is the phenomenon of reciprocal externality. Here, the intertemporal externality is imposed by each well owning farmer on another well owning farmer at a time and over time at a given space. As a consequence, drilling costs increase as the water table falls[1]. The externality affects both the poor and rich people, both spatially and temporally. The phenomenon can be appreciated considering the plight of most of the dug wells due to the advent of deeper borewells, as borewell farmers had better access to groundwater by drilling deeper, when compared with the traditional dugwell farmers[2].

Research on the estimation and impact of externalities in groundwater irrigation in India was first initiated by the Department of Agricultural Economics, University of Agricultural Sciences, Bangalore supported by the Ford Foundation, New Delhi during 1994, by a grant for conducting studies on the interactive effects among irrigation wells[3]. Here, the first study concerning externality due to cumulative interference among irrigation wells entitled 'Equity issues in groundwater development—institutional analysis of failed wells in Karnataka' was conducted between 1994 and 1998. It would be appropriate to appreciate the suggestions provided by Dr John Ambler, the then Program Officer, The Ford Foundation, New Delhi, who recognized the predicament of farmers due to cumulative interference among irrigation wells and sponsored this study at UAS Bangalore. The reasons for well failure were analysed along with how they affect net farm incomes of different categories of farmers in different agroclimatic zones of Karnataka. In the subsequent grant by the Ford Foundation for the second phase of this study entitled 'Dynamics of access to ground water resource in the context of watershed development program', the on farming coping mechanisms (if any) adopted by farmers and the efforts of the government in term of watershed development programs were initiated from 1999 to 2004. It is further appropriate to recollect the contributions of Dr A Ravishankar, Economist at erstwhile NCAP, New Delhi towards measuring contribution of watershed development programs in the proposal which was appreciated by Dr Ujwal Pradhan, the then Program Officer of The Ford Foundation to continue the Ford grant to UAS Bangalore. Both these grants enabled several graduate students to work and contribute towards research on Economics of Groundwater irrigation in hard rock areas under my guidance. Contributions of Professor M V Nadkarni in sampling methodology by including not only functioning wells but also the failed wells of farmers was a significant development. Another notable contribution was by Sri SV Jagannathan, the hydrogeologist at the Central Groundwater Board, Government of India (Southern Region) who helped in developing the index of

[1] Dasgupta 1982

[2] Results (www.toenre.com) have provided ample evidence of the substantial interactive effects of wells and with the inclusion of negative externality costs, the annual amortized cost of irrigation works out to around ₹. 14,000/farm.

[3] http://www.toenre.com/downloads.html list several studies on these aspects covering the first and second phases of the grant by the Ford Foundation.

cumulative interference among irrigation wells, as the (the number of irrigation wells divided by groundwater available in MCM) in each taluk.

Lisa Pfeiffer and Cynthia Lin[4] (2012) reported the first study explicitly reporting on interference among wells in the US. They reported that, whenever, cones of depression overlap, the combined effect on water levels across seasons will equilibrate over the nonpumping season as governed by the hydrogeological characteristics of the aquifer. The groundwater farmers are affected by their neighbours' pumping through overlapping cones of depression in a cropping season, and by the intra-seasonal equilibration of water levels during noncropping season. In both the seasons, withdrawal by one farmer lowers the water table and increases the future pumping costs for neighbouring farmers, shifting the inter-temporal depletion path towards the present.

2.1.2 *Explanation of Externality*

Suppose there are two farmers, a farmer who has drilled shallow borewell, called shallow well farmer (SBF) and the other, deep borewell farmer (DBF). According to Baumol and Oates (1988), an externality is present when SBFs production function includes real variable/s (nonmonetary) whose value is chosen by DBF, without regard to SBFs welfare. First, the action of DBF is unintended or not deliberate and has resulted in reduced yield of groundwater (a side effect) on SBF. Second, there is no compensation (or price system) paid by DBF for the loss suffered by SBF in the process. Third, there is no contract (or rule or institution such as say tit for tat) governing the effect of DBF on SBF. Fourth, this results in over extraction by the DBF which is due to inefficiency caused by negative externality. Fifth, there is welfare loss, since the social costs are more than the private returns from groundwater extraction.

2.1.3 *Why Farmers Overextract Groundwater: Perspective of Negative Externality*

The DBF farmer extracts groundwater (shown on X axis) to the level where his MPC = MPB, at OY^0 (Fig. 2.1). However, DBF ought to have extracted only OY^*, but is actually extracting the excess given by the difference between OY^0 and OY^*. The reason the farmer extracts more groundwater is, because the farmer does not recognize that his or her actions are imposing a cost to the society given by the marginal social cost (MSC; on Y axis). Thus, the DBF overextracts groundwater to the level of OY^0, discounting the MSC. The DBF ignores the cost s/he is imposing on the neighboring SBF, who loses his/her groundwater irrigation well. Thus, the cost of extraction of groundwater for DBF ought to be marginal private cost

[4] http://digitalcommons.unl.edu/cgi/viewcontent.cgi?article=1352&context=usdeptcommercepub.

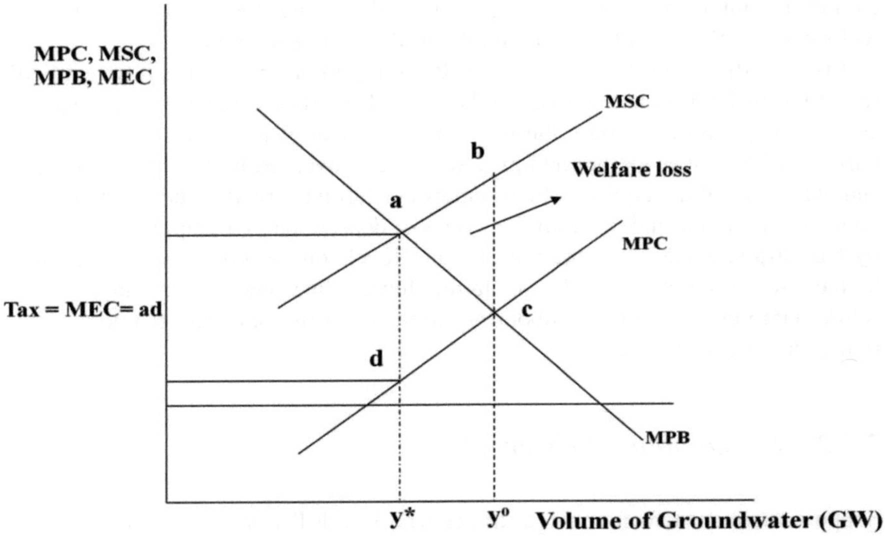

Fig. 2.1 Negative (reciprocal) externality leading to overextraction of groundwater

(MPC) of extraction + cost imposed on the neighboring farmers including themselves over time and on others over space. This results in, inefficiency since DBF goes on increasing his or her extraction, without any concern to SBF. Thus, marginal externality cost (MEC) = MSC−MPC. Since the society does not recognize the MSC, the DBF farmer continues to overextract to the level of OY^0, thus, she or he is extracting OY^0 more than the social optimum of OY*, since MPB = MPC. Thus, DBF and society both ignore this externality or social cost. This results in (i) inefficiency given by over extraction = $y_0 - y^*$ and (ii) welfare loss = abc given by the difference between total social cost and the total private benefit. Since there is no social benefit, private benefit is equal to the social benefit. Accordingly, negative externality results in over extraction of groundwater and welfare loss, as the total private benefit does not cover the total social cost imposed due to overextraction. The way to avoid this overexploitation is to tax the DBF equivalent to MEC, which is the difference between the MSC and the MPC.

Thus, Inefficiency = overextraction = $[y^0 - y^*]$; Welfare loss = $[y^*y^0ca - y^*y^0ba]$ = Area abc

2.1.4 Why Farmers Hesitate to Invest in Groundwater Recharge: Perspective of Positive Externality

A farmer who has taken up (efforts to) groundwater recharge (X axis), is receiving the marginal private benefit (MPB). Thus, his or her private optimal effort to recharge is OY^0 given by the intersection of MPC with MPB (Fig. 2.2). However,

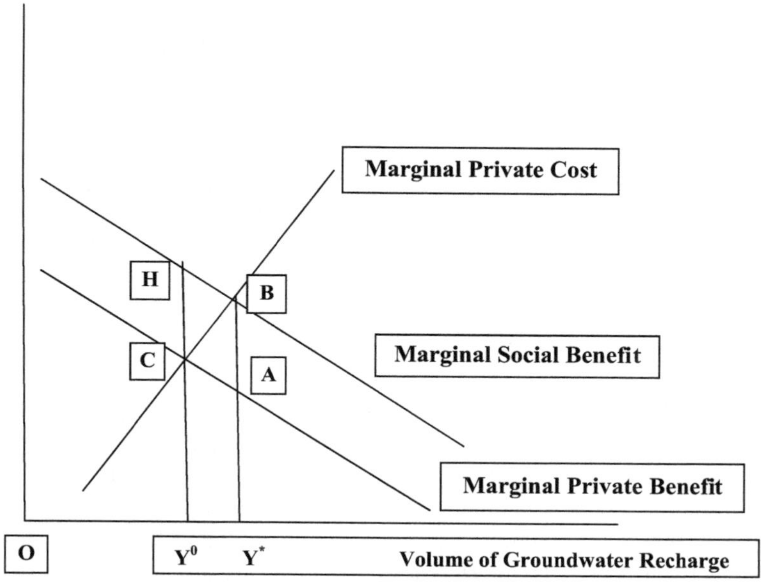

Inefficiency = under utilization of recharge efforts = [OY*- OY^0];
Welfare loss = [Y^0 Y*BH - Y^0 Y*BC = CBH] = Area HBC
Pigouvian Subsidy = AB*(Y*-Y^0)

Fig. 2.2 Positive externality leading to low groundwater recharge effort

the farmer's efforts to recharge groundwater on the farm also recharges the aquifer and results in the societal benefit given by the marginal social benefit (MSB) curve, the shifted curve to the right of MPB curve. Thus, it is desirable for the farmer to undertake recharge to the extent of OY*, but, the farmer is limiting recharge efforts to the extent of OY^0, since neither s/he nor the society does recognize this social benefit. The inefficiency in this process is given by the reduced level of recharge effort OY*−OY^0. The social cost here is equal to the triangle CBH. Here, the total social benefit foregone Y^0 Y*BH is the total social cost because the farmer does not undertake recharge effort to the level of OY*, and hence the society loses the total benefit given by the area under the MSB curve for that extra recharge effort missed (OY*−OY^0). The total cost which is not incurred is Y^0 Y*BC, becomes the total benefit. Thus, the social benefit foregone becomes the total social cost and the total private cost which is not incurred becomes the total private benefit. There is no other cost other than the private cost. Hence, the difference between the two given by Y^0 Y*BH−Y^0 Y*BC = CBH gives the welfare loss. Thus, in order to enthuse the farmer to undertake recharge efforts to the tune of OY*, the farmer has to be subsidized to the tune of the difference between the MSB and MPB = HC. The message thus, is positive externality also leads to inefficiency and welfare loss, which can be corrected by offering subsidy to the tune of the difference between

MSB and MPB. Negative externality leads to inefficiency and welfare loss, which can be corrected by taxing to the tune of the difference between MSC and MPC as shown in Fig. 2.1.

Winner's Curse Farmer, in his or her efforts to remain on the original production possibility curve, invests to maximize access to groundwater. Due to the tendency of over estimation of yield of groundwater at the time of drilling, in order to prove that the borewell point is the right one by the diviner-driller, farmers having invested substantially, tend to lose, after they realize that the well is a low yielder.

2.1.5 Interactive Effects Among Wells

The phenomenon of interactive effect of wells, also known as well interference, refers to the groundwater nexus among wells. Here, among connected wells due to cone of depression, withdrawal of groundwater from one well will result in reduction of groundwater yield and water level in the other. In practice, it is likely that there may be either one well to one well interaction, or one well to many wells multiple interactive effects. The interference phenomena are obscure; to the extent that one or more wells may be causing the problem, therefore it is in order to study the predicament faced by farmers due to 'cumulative' well interference (CWI), since it is difficult to discern the effect of a specific well on another well. By 'cumulative' we mean the sum total effect of overpumping of ground water from several (types and numbers of) wells at a time and over time resulting in the reduction of yield and water levels in the surrounding wells.

The National Geophysical Research Laboratory, Hyderabad, conducted thousands of pump tests and recorded the discharge and water levels for wells with different interwell spacings and recommended that the isolation distance between two borewells should be at least 250 m, to avoid the problem of interference. However, it does not mean that if this distance is maintained, there will/will not be interference. This is a precautionary long-term measure where limits prescribed may help irrigation wells to yield for a long duration of time if water is not overextracted in relation to recharge.

2.1.6 Why Study Interference?

The well interference problem poses serious threats to equity and sustainability in well irrigation. A large farmer who can afford to dig, drill or deepen many irrigation wells can seriously hamper the irrigation prospects of neighbouring (small) farmers irrigating with one or two wells. In the long run, the (small) farmer with borewell irrigation may be forced to shift his or her operations to rainfed farming, if the neighbouring farmer drills deep and overextracts groundwater. There is a clear equity issue here. When the process of withdrawal of groundwater without regard to recharge efforts as also neighboring farmers goes unabated, the resource itself may become unsustainable, clearly reducing the economic life of the well, which underscores the sustainability issue.

2.1.7 How to Study Interference

Cumulative interference among wells is studied through a combination of statistical and participatory rural appraisal (PRA) approaches for selecting farmers to study and analyse the equity issues involved due to interference of irrigation wells. This is because interference is expressed through initial and premature well failure, deepened wells, changed structure of (dug to bore) wells, increased Horsepower of Irrigation pumpsets, adoption of micro irrigation, shift in crop pattern and so on. Hence interference is both visible and invisible, depending upon the researcher's vision. In the first step, the area affected by cumulative interference is chosen using secondary data as under. The secondary data are necessary to identify the taluks/villages suffering from interactive effect(s) of well(s). At the taluk/village level, it is difficult to have a list of taluks/villages which suffer from well failure due to interactive effect(s) of well(s) as pump tests are expensive to be carried out, and hence relevant data are not recorded and available anywhere. Hence, the need to generate a realistic index using the available secondary sources of information. Cumulative interference can best be represented by 'the number of irrigation pumpsets per million cubic feet of utilizable groundwater for irrigation in a taluk'[5]. This index reflects the dependency of a given number of irrigation wells on the utilizable groundwater for irrigation (in an area say taluk or a village). Hence, the taluk with the highest number of wells per unit volume of groundwater reflects high well interference problems compared to a taluk which has the lowest number of wells per unit of groundwater. This index hypothesizes that cumulative interference exists all over (the HRAs) since the irrigation wells are mushrooming, without maintaining the isolation distance. This index is a relative, but not an absolute measure. The data on the number of irrigation pumpsets used to pump groundwater is available from the power utility (say for example KPTCL Karnataka Power Transmission Corporation Ltd in Karnataka). The data on groundwater endowment, i.e. how many million cubic feet of utilizable groundwater is available for each taluk from reports of the Central Groundwater Board. After obtaining the index using the data provided above, the taluk with high or relatively high index of cumulative interference can be chosen. The same procedure can be used to find the cumulative interference at the village level, using similar data sets for the village level. Upon zeroing in the village/can be with relatively high cumulative interference Participatory Rural Appraisal (PRA) can be conducted.

[5] Sri V Jagannathan, Hydrogeologist of the Central Groundwater Board, Government of India devised this index, at a State level seminar held in the Dept of Agri Economics, UAS Bangalore during May 1995.

2.1.8 PRA Approach to Locate the Area of Cumulative Interference Within a Selected Village

After choosing the taluk and the village using the secondary data, farmers owning wells and a few other farmers need to be requested to come to a common location, such as a school or a community hall premises. The farmers can be requested to participate in the PRA mapping exercise. They must be facilitated to draw a base map marking the village boundaries, roads, infrastructure points. Then, crucial variables, such as size of well owning farmers, location of borewell/s in the village, status of each well (whether functioning or failed, failed can be represented by red colour, while functioning by green), year of drilling, year of well failure, interwell distance, seasonal variations in depth of water, yield of the well, area irrigated, capacity of pumpset, cropping particulars, number of wells owned by each farmer, interwell distance intrafarm and interfarm, coping mechanisms adopted by farmers to endure well interference and many other variables which contribute to interference. It is crucial to also obtain the pulse of cumulative interference, by asking farmers to indicate, exactly since which year each farmer felt that his or her well's water yield is /was lowered and also due to which well(s) drilled in whose plot/s at what depth/s. Many may not be able to attribute this clearly to specific well(s), however, will help in obtaining a feel for the degree of reciprocal externality.

2.1.9 PRA Mapping: Wealth of Results on Groundwater Wells, Interference

In a PRA mapping conducted in Bhattarahalli village in Kolar taluk, Karnataka, by MYRADA to map irrigation borewells (July 6–9, 1994. http://myrada.org/myrada/pra9), farmers and local water diviners participated. Information on groundwater exploration diffuses by 'word of mouth' among farmers. Understanding the situation of failed wells from the perspective of farmers is crucial since they are the ultimate field decision makers and investors for coping with the well interference problems. The farmers' perception of failure is different from the technical definition of well failure. According to NABARD, a failed borewell yielded less than 1000 gallons/h. Farmers use different symbols to indicate well failure while drawing the PRA map of well interference on the ground which is translated into a PRA map; Different types of functioning wells are shown in green and failed wells are shown with other marks. Though the PRA map is not according to scale, the area of concentration of failed wells which is in the southern side of village, indicates the area of high interference among wells, when compared with other parts of the village. For detailed studies regarding estimation of externalities, farmers are sampled and responses can be obtained through personal interviews.

2.1.10 Dowsing

Local water diviners played a critical role in locating groundwater using the divining or dowsing (Y shaped) plant or iron rod. Around 60 % of the groundwater wells were located by them and hence wield substantial influence on groundwater exploration by farmers. Hence, farmers believe groundwater diviners and drillers rather than geologists, since there is no effort to educate farmers by hydrogeologists similar to agricultural extension efforts. When the local diviners were asked to explain (i) direction of movement of groundwater, (ii) method used in water divining, and (iii) the hydrogeological formations, their explanation was unscientific and each diviner's perception was distinctly different from others. Farmers and diviners indicated the local methods as under while dowsing groundwater:

1. Locating well where the Y-shaped tree stem (the stick should be lean and wet) turns towards earth, where groundwater is located.
2. Plumb bob method: the ball starts moving or turning around itself when it traces groundwater using the following material:
   i) Magnifying lens and thick copper wire tied to a thread.
   ii) Magnifying lens and precious stone tied to a thread.
   iii) Magnifying lens and metal watch tied to a thread.
3. Carrying dehusked coconut in palm along the farm, with a belief that coconut would shift up gradually at the place where groundwater is traced.
4. Locating well where a child feels giddy! at the point where groundwater is traced.
5. Asking boon from god in a temple.
6. Locating at north-east geomancy (deva moole).
7. Locating according to 'Anjana method' (witchcraft)
8. Head torch method (?)
9. Locating well near termite hill, near the plant *Calotropis gigantica* var.albiflora.
10. Geophysical survey.

The 'Y' stick and 'Plumb bob' were the more often adopted dowsing methods. One diviner opined that the groundwater availability increases as we move from south to north in an area. The diviner mentioned that normal groundwater yielding fractures could be cited at three distinct depths: 75, 145 and 185 ft. This was ratified by the geologist also. Some of the local diviners mentioned that groundwater would be flowing in a canal beneath the rocks underground. This highlighted the misconceptions of diviners who in turn inform farmers with regard to groundwater occurrence. The hardrock system cannot support the existence of underground canals! Some others mentioned that rainfall is the only source for underground water. They also mentioned that a good volume of groundwater can be traced to intersections of north–south and east–west directions on the ground. Some believed that groundwater exists only in fractures and gaps and that groundwater joins a river during its flow.

One of the local diviners who was acknowledged by the geologist as a person with a better understanding of groundwater than the other local diviners, suggested that both 'local' and geophysical methods be used in locating groundwater; 'local' method to locate the point for drilling and geophysical method to estimate the depth and yield. The geologist agreed with few of the opinions of the local diviners and the geologist was taken to a farmer's field. The 'Plumb Bob' local diviner and the geologist interacted in divining a well point. Though the 'Plumb Bob' diviner's location was ratified by the geologist, he attributed the success of the local diviner to a chance factor.

Farmers' perception of failure may be different from the technical definition, as the technical definition of well failure is a sort of overall average and discounts the coping mechanisms adopted by farmers to use whatever available scarce groundwater. A typical study of well interference will explore well failure defined as:

1. Well that dries up because of new well(s) coming in (but not because of decline in rainfall).
2. Well that loses a large degree of its yield because of new well(s) coming in (but not because of decline in rainfall), and
3. Well that is deepened because of new well(s) coming in.

2.2 PRA Results

From the PRA well map, the average distance among all types of wells was found to be 462 ft which was well below the recommended threshold isolation distance of around 850 ft. Nevertheless, the failure of all the dugwells (100 %) and 45 % of borewells are apparent indicators of the cumulative interactive effects in wells.

About 74 % of the wells were borewells and the rest were dugwells. All the dugwells failed to yield water after 20 years. Since the 1980s the use of local, labour intensive water lifting devices such as Kapile, Yetha and Persian wheel have been given up since they can only lift water from shallow depths of dugwells. A majority of the dugwells are now being used for storing groundwater from borewells. Among the borewells, 28 % of the borewells faced initial failure (i.e. they never yielded any water, from the very first day after drilling), 17 % of the borewells faced premature failure (over time), and currently 55 % of the wells are functioning. The first borewell was drilled in 1972 and is functioning without problems till the date of PRA (in 1994). This point was located by a geologist, and is 400 m away from its nearest neighbouring borewell. It was drilled to a depth of 120 ft, while water was struck at 80 ft and yields 9000 gallons/h (GPH). Presently, the average depth of borewells is 207 ft, and the average depth at which water is struck is 117 ft.

The race for borewells could be well understood by considering the case of a farmer who drilled 11 borewells of which only 3 (27 %) are functioning, whereas the remaining suffered initial failure (73 %). The average yield of borewells which failed prematurely was 1750 GPH. The Central Groundwater Board, Government of India, have drilled an observation borewell and an exploratory borewell very close to the village. The local water diviners provided locations for 60 % of the wells and the remaining 40 % were located by geologists. Of the currently working borewells, water diviners showed 50 % of the points and geologists showed the remaining 50 %.

The findings from PRA provide insights regarding the degree of well failure and interactive effects among irrigation wells. In order to understand well failure and the equity implications thereof, PRA mapping (Shyamasundar 1996) of the following features is helpful in analysing what different categories of farmers do when there is well failure, what their economic losses are, and how they cope with the situation: Land holding size of well owning farmers; type and present status of well; seasonal variations in depth of water and yield of the well; type and capacity of pumpset; year of well establishment, well failure and inter-well distance; cropping particulars' area irrigated by each well; number of wells owned by each farmer; sources of finance for well establishment; proportion of income realized from irrigated, unirrigated, food and nonfood crops; coping mechanisms adopted by farmers to endure well interference.

2.2.1 Winner's Curse

The information on groundwater availability is fraught with lack of fundamental information for farmer, diviner, driller, the government and power utility. Since, no farmer knows regarding the likely yield of the borewell as well as the likely age up to which it may serve satisfactorily, farmers make their investment believing the diviner and/or driller. Many a time, driller and diviner will be the same individual. Since the driller makes a huge investment on the rig, his interest is to motivate farmers to drill well/s so that he can recover the investment. The farmer, in his or her efforts to remain on the original production possibility curve, tries to invest as much as possible to maximize access to groundwater. With the usual tendency of over estimation by the diviner-driller, farmers who would have invested heavily, would tend to lose, after the realization that the well is a low or poor yielder. This phenomenon is akin to Winner's curse as in institutional economics enunciated by Thaler (1988)[6].

2.2.2 Role of Groundwater Management

Given the fact that the 'system as a whole is greater than the sum of its spatial parts'[7], groundwater is a 'system' and hence needs to be managed as a 'whole' and

---

[6] http://www.jstor.org/discover/10.2307/1942752?uid=3738256&uid=2&uid=4&sid=21102225674447

[7] Partha Dasgupta (1982, p. 20).

not in 'parts'. Thus, treating groundwater as a 'private property' is disastrous. In the case of groundwater, irrespective of how it is treated—as a common property or private property or open access resource, if it is not sustainably managed, farmer/s freeride(s), and the groundwater resource suffers from excessive use or overextraction in relation to recharge. Thus, management is crucial to control the use of groundwater and avoid the 'tragedy of the commons' either by institutions (rules/ regulations; by establishing charges or quantity restrictions)/markets/technologies/ governance and avoid the 'tragedy of the commons' which results due to unabated overextraction.

References

Baumol WJ, Oates WG (1988) The theory of environmental policy: 2nd ed. Cambridge, Cambridge University Press, New York, pp 17–18

Dasgupta P (1982) The control of resources. Harvard University Press, Cambridge.

Pfeiffer L, Lin C-YC (2012) Groundwater pumping and spatial externalities in agriculture. J Environ Econ Manag 64(1):16–30

Shyamasundar MS (1996) Interplay of markets, externalities, institutions and equity in groundwater development—an economic study in the hard rock areas of Karnataka, Unpublished PhD thesis submitted to the Department of Agricultural Economics, University of Agricultural Sciences, Bangalore

Thaler RH (1988) Anomalies: the winner's curse. J Econ Perspect 2(1):191–202

Chapter 3
Unidirectional and Reciprocal Externality in Irrigation

3.1 Preamble

According to Partha Dasgupta (2008), there are two types of externalities—*unidirectional* and *reciprocal*. In the case of unidirectional externalities, one or a set of agents inflicts or confers an externality on another (or others), and the direction of the externality is in a direction determined by the existing institutions. In the case of reciprocal externalities (REs), each party inflicts an externality on all others, as it happens in unmanaged common property resources. Baumol and Oates (1988) offered the conceptual definition of an externality; studies on its empirical measurement of an externality in the extraction and use of different natural resources were not available. Thus, the first study on the measurement of an externality in groundwater irrigation was attempted in the Department of Agricultural Economics, University of Agricultural Sciences, Bangalore (Chandrakanth and Arun 1997; Arun 1994).

Unidirectional negative externality (UDNE or UDE): Here, the generator of an externality and the victim/s of an externality can be located, and the inefficiency and welfare loss can be estimated. In the case of groundwater, the estimation of unidirectional externality is difficult, expensive and requires great patience. What is reported here is the result of such an effort measuring unidirectional externality in groundwater irrigation.

RE: Here, all the persons generate externalities at a time/over time or both, and the cumulative externality is substantial, as it affects efficiency and results in welfare loss, the simplest example being traffic congestion.

The interference between irrigation wells leads to an externality due to the overlap of the cones of depression of neighbouring wells. As discussed in the previous chapter, the unidirectional externality in the case of interference of irrigation well(s) refers to the reduction in the yield of one irrigation well due to pumping in another irrigation well, which results in gradual failure of one or both of the irrigation wells. Hence initially, due to the countable number of irrigation wells, what begins as an *unidirectional externality* due to the interference between two wells, spreads to cumulative interference among different wells resulting in *RE* over time involving all

M.G. Chandrakanth, *Water Resource Economics,* DOI 10.1007/978-81-322-2479-2_3

the surrounding wells. In case of RE, it is difficult to identify or single out the specific well or group of wells responsible for interfering with another specific well or group of wells, since there is no 'one to one interference' similar to point pollution, which can be identified, instead, there is 'cumulative interference' similar to 'non-point pollution'. In RE, due to the rat race for groundwater, there is mushrooming of wells, extracting water at unsustainable rates, that is, extracting more than what is recharged similar to the Wantrupian 'fastest and the mostest' concept.

As the first step in research, 'unidirectional' externality was identified. The information regarding unidirectional externality, where one farmer's drilling deep or extracting more of groundwater affects the neighbouring farmer, or vice versa is not recorded anywhere and hence not available. The only way of obtaining such information is through 'snowball sampling'. Here, the first farmer whose well is interfered is identified with the help of a local water diviner. Later, the first farmer is asked to give the name and location of the next farmer whom she/he knows is affected by well interference. After obtaining data from the second farmer, the address of the third farmer is similarly obtained. This procedure is continued till the desired number of sample farmers is obtained. For the empirical study in order to obtain the sample of 40 farmers, 20 villages located in four taluks of Bangalore rural and Kolar districts had to be contacted. This shows that during 1996, interference was not relatively apparent; hence, in order to obtain a small sample of 40 farmers whose wells are interfered, the geographical spread was wide, since the researcher had to cover two districts and four taluks. Thus, locating the sample units in the case of unidirectional externality becomes an expensive and difficult proposition job.

The field data were obtained by a primary survey through personal interviews for the situation before the interference (as pre-well interference) and after the interference (post-well interference) periods. As the farmer/s find it difficult to understand, assimilate and share their opinion on 'externality', the data regarding the crop pattern and net returns obtained in the years of pre-well and post-well interference were obtained to devise ratio variables and hypotheses (Table 3.1).

3.2 Explanation

1. RGAC: Ratio of gross irrigated area under cereals in the post-interference period to gross area irrigated under all crops in the pre-interference period.
2. RGAV: Ratio of gross irrigated area under vegetables in the post-interference period to gross area irrigated under all crops in the pre-interference period.
3. RGAM: Ratio of gross irrigated area under mulberry in the post-interference period to gross area irrigated under all crops in the pre-interference period.
4. RGPV: Ratio of gross profit from vegetable crops in the post-interference period to gross profit from all crops in the pre-interference period.
5. RGPM: Ratio of gross profit from mulberry in the post-interference period to gross profit from all crops in the pre-interference period.

Table 3.1 Ratio variables and hypotheses concerning negative externality in well irrigation in hard rock areas (HRAs)

Variables	Hypothesized relationship	Rationale
Dependent		
Externality = farmer who invested on additional well(s) in the event of well failure due to UDE (1, 0 otherwise)	Farmer drills additional well in the event of well failure due to interference externality	Farmer gambles with groundwater resource by investing on additional well(s) after the existing well failed, using mini–max criterion, as the farmer wants to sustain pre-interference economic status, intending to minimize the maximum loss due to interference
Explanatory		
1. Size of holding (SIZE)	SIZE has a positive influence on the investment in the groundwater resource	Investment on a groundwater well is a strategic management decision governed by the SIZE. There is a threshold SIZE which attracts investment on additional wells, below which an investment on additional wells does not hold good
2. RGAC 3. RGAV 4. RGAM 5. RGPV 6. RGPM	As the ratio tends to unity, the probability of drilling additional wells is lower; hence, there is an inverse relation between the ratio variable and the investment on additional wells	If the post-interference profit or area irrigated is close to the pre-interference profit or area irrigated, then the ratio approaches unity. Thus, the farmer does not invest in additional wells as she/he remains on the same pre-interference production possibility curve (PPC). If the ratio is close to zero, then post-interference profit or area irrigated is far lower than the pre-interference profit or area irrigated. This motivates to invest in additional wells to enable the farmer to be on the pre-interference PPC
7. RAAF	As the ratio tends to unity, the probability of drilling additional wells falls	The farmer tries her/his best not to lose the extent of area irrigated by wells. Higher RAAF implies higher intensity of the effect of well interference which motivates the farmer to invest in additional wells
8. RAWY	As the ratio tends to unity, the probability of drilling additional wells falls	If the ratio of groundwater yield in the post-interference period to the pre-interference period tends to unity, no reduction in groundwater yield is implied due to interference, and hence the farmer does not invest in additional wells

Table 3.2 Logit probability of drilling an additional well by farmers, given well failure

Variables	Logit coefficient	*t*-value	Elasticity of probability
1. Size of the holding (SIZE)	0.38	2.62*	0.35
2. RGAC	−0.52	−0.19	0.26
3. RGAV	11.63	2.32*	
4. RGAM	−6.20	−2.69**	0.31
5. RGPV	0.85	0.62	
6. RGPM	0.35	1.23	
7. RAAF	8.62	1.82*	0.06
8. RAWY	2.38	0.597	
Constant	−8.37	−1.74*	
Log likelihood ratio test		24.85*	
Conditional probability of drilling an additional well (P)		0.87	
Mc Fadden R^2		0.48	

* Significant at 5 %; ** significant at 1 %

6. RAAF: Ratio of net irrigated area affected by well interference to total irrigated area.
7. RAWY: Ratio of average water yield of the interfered well to average water yield of the well in the pre-interference period.

In this study, the probability that a farmer would invest in an additional well reflects the unidirectional negative externality (UDNE) in well irrigation due to interference. From among the 40 sample farmers who are facing the UDNE (treated as a dependent variable), those farmers who have invested in an additional well (assigned dummy variable value of 1) in the event of well failure due to UDNE are dichotomously differentiated from those who did not invest (dummy variable value of 0), using the binomial logit model (Table 3.2). The results indicated that the probability of UDNE significantly increased with holding size, relative area under vegetables (RGAV) and the relative area affected by interference (RAAF) and reduced with the increase in relative area under mulberry (RGAM).

3.3 Cultivation of Water-Intensive Crops Enhances Negative Externality

The overall probability of UDNE due to interference is estimated at 0.87 (Table 3.2), which gives an odds ratio of 6.69 (=0.87/0.13). For every one chance of nonoccurrence of UDNE, there are 7 chances of occurrence of UDNE. If the size of holding increases by 1 %, the probability of UDNE increases by 0.35 %. If the ratio of gross irrigated area under vegetables in the post-interference period to gross area irrigated under all crops in the pre-interference period increases by 1 %, the probability of UDNE increases by 0.26 %. If the ratio of net irrigated area affected by well inter-

Table 3.3 Estimation of willingness to pay for an additional well (Tobit)

Variables	Tobit maximum likelihood estimator coefficient	*t*-value	Conditional marginal willingness to pay for an additional well
1. Size of the holding	14,122	3.57**	14,122
2. RGAC	−21,247	−0.25	−212
3. RGAV	1,47,680	1.57*	1477
4. RGPV	−23,003	−0.74	−230
5. RGAM	29,885	0.51	209
6. RGPM	9059	1.04	91
7. RAAF	61,412	0.60	614
8. RAWY	11,435	0.12	114
Constant	−168680	−1.49*	–
Estimated unidirectional negative externality (₹)	48,370		

*Significant at 5%, **Significant at 1%

ference to total irrigated area increases by 1%, the probability of UDNE increases by 0.06%. If the ratio of gross profit from mulberry in the post-interference period to gross profit from all crops in the pre-interference period increases by 1%, the probability of UDNE decreases by 0.31%. The results are clear indicators of the factors which increase/decrease the negative externality. While the area under vegetables increases the negative externality, the area under mulberry decreases the negative externality because vegetables are water-intensive crops, compared with perennial low-water-intensive crops such as mulberry. Similarly, an increase in profits also contributes to negative externality since it is due to 'beneficial' use of groundwater rather than 'wise' use.

3.4 The Measure of UDNE

With the finding that the probability of drilling an additional well due to the failure of existing well(s) and due to UDNE, farmers can drill an additional well or maintain the status quo. The investment of an additional well will be the empirical measure of UDNE. Using the Tobit model for the same set of 40 farmers and the investment made by farmers (which in itself is the UDNE) in the event of well failure as the dependent variable, with the same set of explanatory variables as used for the logit model above, the estimated UDNE was ₹ 48,370. For every 1 acre increase in the size of the holding, the farmer would invest further ₹ 14,122 which further increases the UDNE (Table 3.3). Despite the high probability rate of well failure in the hard rock areas (HRAs) (=0.4), this reflects the capture of the resource in spite of uncertainty.

Considering the size of holding as an economic activity which determines the farmer's ability to invest on a new well in the event of well failure, the reservation holding size which prompts the farmer to invest in an additional well is found to be 5.35 acres. This shows that farmer should at least have 5.35 acres of holding size, to invest in a new well in the event of failure.

Unidirectional externalities in well irrigation in HRAs are intricate. With the high probability of drilling an additional well of 0.87 and the high probability of well failure of 0.40, the predicament of negative externality gets exacerbated, as farmers are involved in both causing and bearing the brunt of groundwater overdraft. Farmers further add to negative externality by investing on an additional well instead of considering of other demand side factors such as using drip irrigation, for instance. Thus, with the reducing life and age of the well(s) in the HRAs, despite the free cost of power to lift groundwater, the cost of groundwater shoots up due to the investment on additional well/s and due to premature well failure. This calls for an economically sound costing methodology of well irrigation in HRAs.

References

Arun V (1994) Estimation and internalization of externalities in groundwater irrigation in hard rock areas of Karnataka. Unpublished M. Sc (Agri) thesis, submitted to the Department of Agricultural Economics, University of Agricultural Sciences, Bangalore

Baumol WJ, Oates WG (1988) The theory of environmental policy, 2nd ed. Cambridge University Press, Cambridge, pp 17–18

Chandrakanth MG, Arun V (1997) Externalities in groundwater irrigation in hard rock areas. Indian J Agric Econ 52(4):761–771

Dasgupta P (2008) The place of nature in economic development. The Beijer Institute of Ecological Economics, Beijer Discussion Paper Series 216, Royal Swedish Academy of Sciences, Sweden, p 14

Chapter 4
Sand Mining Externality

4.1 Encroachment

As long as the river flows to its natural length and breadth, there would be no scope for farmers to encroach on the riverbeds. However, due to erratic rainfall, the dimension of river flow reduced, exposing the layers of sand. Riparian farmers gradually began to encroach on the riverbeds on either side of the river, which are a state property (not a common property). Further, farmers began cultivating these fertile lands by drilling filter point wells and began extracting groundwater for irrigation since 1980s.

As the demand for sand increased, the riverbeds began to be exploited. With the worsening of rainfall, river flow as well as the recharge was affected. Accordingly, the filter point wells drilled in the riverbed, open wells and borewells constructed along the riparian were affected. The harvest of sand continued unabated resulting in virtual mining of sand, imposing negative externalities on sand mining farmers as well as other farmers in the riparian areas.

The Department of Mines and Geology (Karnataka) issues a permit license for sand mining to contractors to extract sand from the top 3 ft. In practice, the contractors extract exploitatively as the sand prices increase and there is great demand for sand in construction and repair work. A majority of the riparian farmers have themselves been involved in extracting sand since (1) they singlehandedly could not halt the sand mining and (2) income from sand extraction was more remunerative than from agriculture. Farmers and contractors were hand-in-glove with opportunistic behaviour, realizing disproportionately high incomes unmindful of the long-term effect of sand mining. The sand is being extracted on a first come, first serve basis.

M.G. Chandrakanth, *Water Resource Economics*, DOI 10.1007/978-81-322-2479-2_4

4.2 Overlap of Externalities

Sand contractors and riparian farmers are mining the sand in a few riverbeds that have unidirectional externality on groundwater levels in the open wells, filter point wells and borewells. When sand, the source of recharge for groundwater is mined, the groundwater level in wells in the riparian areas is affected, resulting in initial and premature failure of wells. This, in addition to already existing reciprocal externalities, is exacerbating the predicament due to overlap of unidirectional externality from sand mining with reciprocal externality of drilling wells.

4.3 Sand Mining in Palar Pennar Basin

Sand mining is carried out along the North Pinakini, South Pinakini and Palar river basins. The main source of sand is the North Pinakini basin, where around 700 truckloads of sand are extracted and transported, followed by the Palar basin (150 lorry loads per day) and Dakshina Pinakini (50 lorry loads per day) (Table 4.1).

In this chapter, the environmental and economic effect of sand mining on groundwater resources and its impact on agriculture are estimated for policy implications. The data pertain to Gauribidanur taluk along the Uttara Pinakini river where sand mining is prevalent. Sand mining is reducing the recharging capacity of groundwater in riparian areas and is reflected in increased probability of well failure. This is resulting in crop losses due to reduced irrigation water in the wells. The unidirectional negative externality due to sand mining along with the reciprocal negative externality is reducing farmers' incomes substantially.

4.4 Unidirectional Negative Externality Leading to Overextraction of Sand from River Streams

Sand extractors, while extracting sand more than the legal volume, are inflicting negative externality on the riparian farmers who are irrigating their lands from irrigation wells, recharged by layers of sand deposited in the river stream. The layers of sand have a dual role to play: (i) of supporting the (mangrove) vegetation which protects the riverbeds and hence the length and breadth of river flow and (ii) of recharging groundwater as sand particles hold water and also allow lateral and vertical flow of water to neighbouring well/s.

Thus, sand mining affects farmers dependent on wells irrigating their lands. However, the sand miners bear the marginal cost of sand extraction, and transfer the negative externality on riparian farmers. Thus, the sand miners bear only marginal private cost (MPC) of extraction, while forgetting that they are also imposing a social cost due to mining. In addition, the sand miners are also imposing the marginal social

Table 4.1 Extent of sand mining along the rivers in Kolar District. (Source: Writ Petition No. 35946 (2001), High Court of Karnataka, Bangalore)

Sl. No.	Taluks	Rivers and streams	River length (Kms)	Width of river stream (meters)	Depth of sand (meters)	Volume of sand accumulated in thousand metric tonnes	Sand transported per day in thousand metric tonnes
1	Gauribidanur	Uttara Pinakini and	58	100	3	17400	2.65
		Kumudvathi	56	40	2	4488	
	Gudibande	Kushubhavathi	43	30	3	3870	1.77
	Bagepalli	Chitravathi	48	60	3	8640	
	Chikkaballapur	Vandamaan	24	25	2	1200	1.77
	Chintamani	Papagni	68	70	3	1428	
A	Uttara pinakini and other sub-streams		296			49878	6.19
2	Chikkaballapur	Dakshina Pinakini	38	30	2	2280	0.44
	Shidlagatta	Markandaya	50	30	2	3000	
	Bangarpet	vrishibavathi	58	40	2	4640	
	Malur						
B	Dakshina Pinakini and other sub-streams		146			9920	0.44
3	Kolar	Palar	108	70	3	22680	1.32
	Srinivasapura	Kundanya	28	20	2	1120	
	Mulubagalu	Kushavathi	38	20	1	760	
	Bangarapet						
C	Palar and other sub-streams		617			24560	1.32

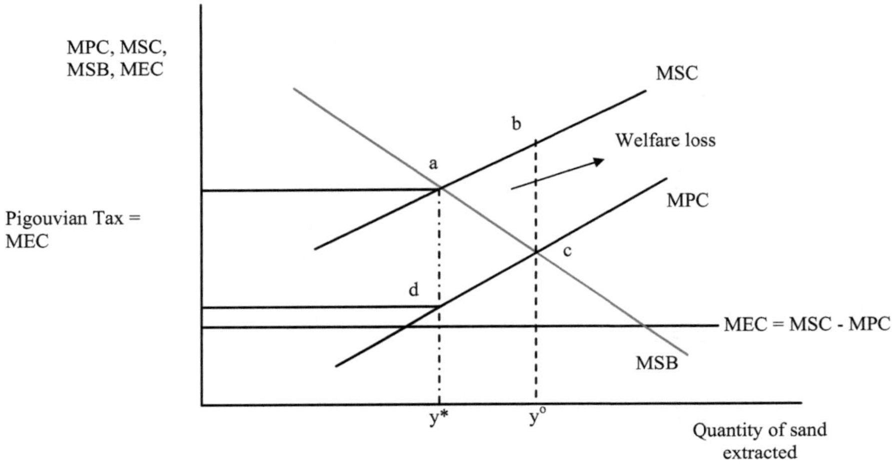

Fig. 4.1 Reciprocal negative externality due to sand mining in Uttara Pinakini river basin of Karnataka

cost (MSC) due to negative externality on riparian farmers. The negative externality imposed by sand mining is given by the marginal externality cost MEC = MSC − MPC. Since the sand miner optimizes sand extraction at the point of intersection of MPC and MSB, he/she is extracting y^{o} quantity of sand. But, considering the MEC, sand miner should extract the socially optimal extraction of sand y^* which is lower than the private optimal quantity y^{o}. However, both the farmer and society ignore the MSC. This misallocation of resources is a market failure and results in a welfare loss equal to triangle abc as well as inefficiency to the tune of excess extraction of $y^{o} - y^*$. In order to make the sand extractor extract the socially optimal sand y^*, the society needs to impose a Pigouvian tax on each tonne of sand extracted by the sand extractor, above the level y^* such that his/her MPC = MSC = ad (Fig. 4.1).

4.5 Unidirectional Externality

Riparian farmers located on either side of the Uttara Pinakini river stream extract groundwater for irrigation from filter point wells, borewells and open wells. The stream has been virtually dry without any river water since the past few years. Farmers chosen for this study are those located within the distance of around 1200 ft from the Uttara Pinakini river stream in Gauribidanur taluk, in Kolar district. It is hypothesized that sand extraction up to 3 ft from the surface does not significantly influence the horizontal and vertical recharge of groundwater and the groundwater yield of wells, according to the rule of sand extraction permitted by the Department of Mines and Geology. Thus, sand extraction beyond 3 ft in the river stream is hypothesized to inflict negative externality by affecting the horizontal and vertical recharge of groundwater. According to Partha Dasgupta (1982), this externality is classified as unidirectional externality, since sand mining affects farmers with irrigation wells and not vice versa. Sand mining activity thus affects groundwater recharge that was earlier facilitated by different layers of sand acting as the spongy layer.

4.6 Reciprocal Externality

When sand mining becomes intense, resulting in sand excavation beyond 3 ft from the surface, filter point wells, open wells and borewells are prone to initial/premature failure. This would result in farmers drilling additional borewells farther and deeper in response to a lowered groundwater table due to sand mining. Due to the failure of filter point wells and open wells, farmers drilled borewells, and overtime cumulative interference among the borewells resulted in well failure. Thus, in the sand mining situation, there is a simultaneous occurrence of unidirectional externality from sand mining and reciprocal externality due to cumulative interference from irrigation wells. Thus, investment on irrigation wells in this situation subsumes unidirectional and reciprocal externality.

4.7 Amortization of Investment

Investment in irrigation wells in this situation is amortized over the average life of all (functioning and non-functioning) wells. Thus, even though investment on irrigation wells is treated as a fixed (sunken) cost, the imminent negative externality arising from the sand mining activity shortens the life and age of irrigation wells, and in turn results in additional investment on either deepening existing irrigation wells or drilling new wells. The amortized cost is hence a variable cost and enters the production function as the cost of irrigation if the farmer internalizes it. Otherwise, it remains as a component of negative externality.

4.8 Encroachment of Riverbeds and Failure of Filter Point Wells

When rainfall was sumptuous and adequate, the Uttara Pinakini river was full, enabling farmers to cultivate their land in the riparian areas. As rainfall distribution is becoming erratic over the years, the river course and/or river width is narrowing, exposing the sanded portion of the riverbed. With reduced and virtually no flow of water in the river stream, farmers on either side of the river bank have began to encroach on the riverbeds for cultivation over time. In search of well water, farmers

Table 4.2 Depth of sand mining, sample size, in Pinakini basin in Gauribidanur, Karnataka

Sl. No.	Particulars	Sample farmers				Sand suppliers (30)
		Marginal	Small	Large	Total	
1	Sand mining area (SMA): Kalludi, Heribidanur: (Mining 5–8 ft)	9 (23)	19 (47)	12 (30)	40 (100)	1–4 trucks/day
	Veeralagollahalli, Doddakurubarahalli: (Mining 10–30 ft)					25 trucks/day
2	Non-sand mining area (NSMA) (Chigatagere, Nandhiganahalli, Chikkakurugod, Mavenakanahalli)	2 (7)	18 (60)	10 (33)	30 (100)	0

Figures in parentheses are percentage to the respective total

drilled filter point wells in the encroached riverbeds on either bank of the river. Due to sand mining, the filter point wells are also failing. Filter point wells are highly sensitive for recharge and discharge. They get quickly recouped due to rains as the water does not have to infilter through hard rocks.

As filter point wells were vulnerable to failure, this hints at volumes of negative externality and economic loss to farmers, even though they have no property rights on the encroached land. In this study, farmers' investment on filter point wells and other wells in the riverbed is considered as an investment, though farmers are not authorized to drill wells in the riverbed. Thus, the riverbed that was a common property resource earlier for water and sand has now been converted into a private property resource.

4.9 Sample

The influence of sand mining is reflected in the open wells, borewells and filter point wells which supply irrigation water to farms that are in proximity to the sand mining area (SMA). In order to capture the influence of sand mining and compare the same with normal riparian habitat farms, the sample consists of farms that are drawn from areas adjacent to stream from both the SMA and the non-sand mining area (NSMA). For this study, a sample of 40 and 30 farmers have been surveyed in the SMA and NSMA, respectively, to estimate the externality. In order to estimate the demand function for sand, a sample of 30 sand truck sellers in Bangalore city (Hebbal, Yelahanka, Yeshwanthpura, Vidyaranyapura and Gangenahalli) are interviewed[1] (Table 4.2).

The Department of Mines and Geology conducted a preliminary survey before leasing the area for mining along the Uttara Pinakini river to regulate depth of min-

[1] Hemalatha 2003.

Table 4.3 Land use pattern and net income in sand mining and non-sand mining

Type of land	Non-sand mining area $n = 30$			Sand mining area $n = 40$		
	Size of land holdings (acres)	Gross cropped area (acres)	Net return per acre (₹)	Size of land holdings (acres)	Gross cropped area (acres)	Net return per acre (₹)
Rainfed	44 (29)	23 (13)	3880	49(19)	27.5(18)	2380
Irrigated	98 (62)	152 (87)	12316	157(61)	127(82)	9992
Fallow	15 (9)	NA	NA	53(20)	NA	NA
Total	157 (100)	175 (100)	NA	259(100)	154.5(100)	NA
Average	5.2	5.8	12492	6.4	3.8	8639

NA not applicable

ing on a sustainable basis. Accordingly, it was decided to permit sand extraction up to 3 ft. However, this permission was misutilized following the perennial demand for sand due to civil works constantly undertaken in the nearest megacity, Bangalore, resulting in sand mining beyond permissible limits. In Kalludi and Herebidanur villages, sand is being mined beyond 12 ft. This has even reduced the availability of good quality sand. As a result, a mere 2–3 truckloads of sand are mined every day. However, in Doddakurubarahalli and Veeralagollahalli villages, the depth of sand mining is up to 25 ft and about 20–25 truckloads are mined every day, as the quality of sand is desirable in this area and fetches a relatively higher price. Due to a good transportation facility and labour availability in these villages, sand is being extracted in an unsustainable manner resulting in an adverse impact on the riparian environment.

4.10 Land Use Pattern

In SMA, around 82 % of the gross cropped area is irrigated and net returns per acre were ₹ 9995, whereas in NSMA, around 87 % of gross cropped area is irrigated and net returns per acre were ₹ 12,316. The decreased net returns per acre in SMA are attributed to increased cost of irrigation due to externality on groundwater extraction, as groundwater yields have declined in all the three types of irrigation wells (Table 4.3).

In SMA, 20 % of the total land holdings have been left fallow as against 9 % in NSMA. This is because farmers are getting an additional source of income from sand mining activity that is resulting in initial and premature failure of borewells, filter point wells and open wells in SMAs (Table 4.4).

Table 4.4 Age and life of irrigation wells in sand mining and non-sand mining areas of Gauribidanur taluk along Uttara Pinakini river, Kolar district, Karnataka

Type of wells	Sand mining area Sample size: 40 farmers			Non-sand mining area Sample size: 30 farmers		
	Number of wells	Average age (years)	Average life (years)	Number of wells	Average age (years)	Average life (years)
Functional borewells	33(79)	10.25	NR	15 (88)	11.28	NR
Non-functional borewells	9(21)	NR	5.5	2 (12)	NR	12
Overall	42(100)			17(100)		
Functional filter point wells	8(32)	7	NR	17(77)	14.23	NR
Non-functional filter point wells	17(68)	NR	19.3	5 (23)	NR	20.66
Overall	25(100)			22(100)		
Functional open wells	0(0)	0	NR	7 (44)	22.75	NR
Non-functional open wells	10(100)	NR	15.76	9 (56)	NR	16.66
Overall	10(100)	NR	NR	16(100)	NR	NR
Total functional wells	41(54)	NR	NR	39(71)	NR	NR
Total non-functional wells	36 (46)	NR	NR	16 (29)	NR	NR
Overall	77 (100)	8.5	14.86	55 (100)	14.62	17.32

Figures in parentheses are the percentage to the respective total
NR not relevant
• Average age of wells refers to age of wells still functioning at the time of field data collection (June 2003). Thus, age of wells = year 2003 minus year of construction of well
• Average life of wells refers to life of wells, which either had initial or premature failure by the time of field data collection (June 2003). Thus, life of wells = Year of failure minus year of construction of well
• Obviously average age is computed only for functioning wells and average life is computed only for failed wells

4.11 Influence of Sand Mining on Well Failure, Age and Life of Irrigation Wells

In SMA, there is a high rate of failure for irrigation wells (46 %) compared with NSMA (29 %). This is a *prima facie* indicator of the effect of sand mining on groundwater resources. All the open wells in SMA failed after functioning for 16

Table 4.5 Cropping pattern, area and net income under open well (near streams) in NSMAs and SMAs along the Uttara Pinakini river basin in Gauribidanur taluk, Karnataka

Sl. No.	Crop	Non-sand mining area		Sand mining area	
		GCA (acres)	NR per acre (₹)	GCA (acres)	NR per acre (₹)
1	Maize	11.5 (39)	4830	All open wells have dried in the sand mining area	
2	Gherkin	0.5(2)	30000		
3	Groundnut	2 (7)	7970		
4	Tomato	3 (10)	30826		
5	Paddy	4 (13)	7115		
6	Sunflower	5 (17)	8020		
7	Sugarcane	4 (7)	18000		
8	Marigold	0.5 (2)	4400		
	Total	30.5(100)			
	Average		10605		

Figures in parentheses are the percentage to the respective total
GCA gross cropped area, *NR* net returns

years as most of the wells were dug to a depth of 30 ft while, on the contrary, sand was being mined to a depth of 30 ft. This severely affected the groundwater level in open wells, resulting in total failure of open wells in SMAs. However, in the NSMA, 44 % of the open wells were still functioning since the past 23 years, and this is another indicator of the vulnerability of irrigation wells to sand mining activity.

Considering the filter point wells in SMA, 68 % of the filter point wells failed to yield groundwater, after functioning for 17 years. In NSMA, 77 % of the filter point wells were functioning since the past 14 years, since 23 % of the filter point wells failed after functioning for 21 years. This also is another indicator of the effect of sand mining on filter point wells. In SMA, the rate of failure (21 %) of borewells was higher compared with NSMA (12 %). The average age of borewells was 38 % lower in sand mining compared with non-sand mining due to adverse impact of sand mining on groundwater resources.

Overall, the NSMA had a much greater proportion of functioning wells of all types, viz. open wells, filter point wells and borewells. (Figure 4.1) In SMAs, the main source of irrigation was borewells (54 %), whereas in NSMAs, the main source of irrigation was filter point wells (44 %) (Table 4.4).

4.12 Cropping Pattern and Net Income

In NSMA, 44 % of the open wells were functioning and gross cropped area under open wells was 30.5 acres, whereas in SMA all the open wells were non-functional due to a dip in groundwater level due to sand mining (Table 4.5).

Table 4.6 Cropping pattern, area and net income under filter point well (near streams) in NSMAs and SMAs along Uttara Pinakini river basin in Gauribidanur taluk, Karnataka

		Sand mining area		Non-sand mining area		Difference in NR/acre (₹)	Per cent difference of NR/acre (₹)
Sl. No.	Crop	GCA (acres)	NR per acre (₹)	GCA (acres)	NR per acre (₹)		
1	Maize	8.5 (48)	4884	17 (38)	6118	1234	20
2	Chillies	0.5 (2)	5380	3 (7)	14936	9556	64
3	Gherkin	0 (0)	0	1(43)	28000	NA	NA
4	Tomato	0 (0)	0	7 (16)	7975	NA	NA
5	Paddy	2 (10)	1975	2.5 (5)	3852	1877	49
6	Sunflower	3 (16)	7333	1(2)	12000	4667	39
7	Mulberry	6 (16)	19500	10(11)	28842	9342	32
8	Sugarcane	0 (0)	0	4(4)	24350	NA	NA
9	Banana	4 (10)	15300	2(2)	23500	22700	59
10	Marigold	0 (0)	0	4.5 (10)	10796	NA	NA
	Total	24(100)		52 (100)		0	
	Average NR/acre		6635		14147	7512	53

Figures in parentheses are the percentage to the respective total
NA not applicable, *GCA* gross cropped area, *NR* net returns

Considering the filter point wells, net returns per acre in NSMA are 53 % higher compared with SMA. The net returns per acre are higher among all the crops in NSMAs. In crops like chillies, banana, paddy and sugarcane, the net returns per acre are higher by 64, 59, 49 and 32 %, respectively, in NSMA compared with SMA (Table 4.6).

With borewell irrigation, in NSMA, around 50 % of the area is under crops maize (34 %) and sugarcane (17 %). In SMA, 28 % is under maize, 20 % under mulberry and 12 % under banana. However, net returns per acre in SMA (₹ 9901) are lower by 15 % compared with NSMA (₹ 11684). This is due to high investment on irrigation wells as a consequent of higher rate of initial and premature failure of wells resulting in higher amortized cost of irrigation. In SMA, net return per acre is higher by 59 % in maize, which is relatively a low water consuming crop. In NSMAs, for high water consuming crops such as sugarcane and banana, net returns per acre are higher by 34 and 28 %, respectively (Table 4.7).

The average net returns per acre are 58 % higher in NSMA for crops cultivated under filter point wells, whereas for crops cultivated under borewells, the net returns per acre are higher by 8 % in NSMA as compared with SMA. This indicated that impact of sand mining is apparent in filter point wells as they are in proximity to SMA and they are of shallow depth (30–35 ft). These wells are obviously the most susceptible for sand mining done to a depth of 25–30 ft. Thus, groundwater in the riparian filter point wells dried up faster compared with borewells located away from SMA drilled to greater depths of 400–650 ft.

Considering the overall performance of SMAs and NSMAs, the net return per acre is 18 % higher in NSMA compared with SMA. The net returns per acre consid-

Table 4.7 Cropping pattern, area and net income under borewell (located away from streams) in NSMAs and SMAs along the Uttara Pinakini river basin in Gauribidanur taluk, Karnataka

Sl. No.	Crop	Sand mining area		Non-sand mining area		Difference in NR/acre (₹)	Per cent difference of NR/acre (Rs)
		GCA (acres)	NR per acre (₹)	GCA (acres)	NR per acre (₹)		
1	Maize	29(36)	5646	19(33)	3652	−1994	−54
2	Chillies	4.5(5)	7033	4 (7)	6332	−701	−11
3	Gherkin	0(0)	0	0.5 (1)	30600	33600	NA
4	Ragi	4 (5)	1767	3.5 (6)	1732	−35	−2
5	Groundnut	0 (0)	0	4 (7)	5432	5432	NA
6	Tomato	0 (0)	0	2 (4)	11200	11200	NA
7	Paddy	6 (7)	2771	6 (10)	5137	2366	46
8	Sunflower	12.5 (15)	7100	4.5 (8)	6700	−400	−6
9	Mulberry	21 (13)	17540	5 (4)	19200	1600	8
10	Sugarcane	8 (5)	15500	19 (17)	23500	8000	34
11	Banana	18 (11)	17206	2(2)	24200	6994	28
	Total	103(100)		69.5(100)			
Net return per acre from borewell			10775		11684	909	8

Figures in parentheses are the percentage to the respective total
For mulberry, sugarcane and banana, the gross cropped area (GCA) is taken as twice the net cropped area, since the crop stays for more than one season
NA not applicable, *GCA* gross cropped area, *NR* net return

ering all types of irrigation wells are ₹ 12,316 in NSMAs compared with ₹ 9992 per acre in the SMAs. This is due to a decline in availability of groundwater in the SMA compared with NSMA (Table 4.8).

4.13 Net Returns per Acre under Different Well Regimes Considering the Proportion of Well Success and Average Age of Well

The impact of sand mining activity, which is reflected through increased cost of irrigation, leads to decreased net returns per acre. The net returns per acre among filter point wells declined by 53 % in SMA (₹ 6635) compared with NSMA (₹ 14,145). After considering the proportion of well success the net returns per acre among filter point wells declined by 79 % in SMA (₹ 2123) compared with NSMA (₹ 10,327). Further, by discounting net returns per acre among filter point wells considering the average age of wells, the net returns per acre declined by 89 % in SMA (₹ 13,740) compared with NSMA (₹ 125,027) (Tables 4.9 and 4.10).

The net returns per acre under borewells declined by 7 % in SMA (₹ 10,775) compared with NSMA (₹ 11,684), whereas after considering the proportion of well success the net returns per acre among borewells declined by 11 % in SMAs

Table 4.8 Cropping pattern, area and net income from Agriculture in NSMAs and SMAs along the Uttara Pinakini river basin in Gauribidanur taluk, Karnataka

Crop	Sand mining area			Non-sand mining area			Per cent difference in NR over non-sand mining
	GCA (Acres)	Cost of cultivation per acre	NR per acre (Rs)	GCA (acres)	Cost of cultivation per acre	NR per acre (Rs)	
Maize	37.5 (30)	7058	5473	47.5 (31)	7061	4820	−13
Chilly	5 (4)	11228	6868	7 (4)	6339	10019	31
Cucumber	0 (0)	–	0	2 (1.5)	–	30500	NA
Ragi	4 (3)	3910	1767	3.5 (2.5)	3520	1732	−2
Groundnut	0 (0)	–	0	6 (4.5)	5460	6278	NA
Tomato	0 (0)	–	0	12 (8)	18970	14225	NA
Paddy	8 (6)	11248	2572	12.5 (8.5)	7945	5513	53
Sunflower	15.5 (12)	10335	7145	10.5 (7)	8876	7833	8
Mulberry	27 (21)	24090	18670	15 (10)	20040	25628	27
Sugarcane	8 (6)	21067	15500	27 (18)	16401	22811	32
Banana	22 (17)	38460	15468	4 (3)	33974	23850	35
Marigold	0 (0)	–	0	3 (2)	4663	10156	NA
Total	127 (100)			152 (100)			
Average			9992			12316	18

Figures in parentheses are the percentage to the respective total
NA not applicable, *GCA* gross cropped area, *NR* net returns

Table 4.9 Net returns per acre under different well regimes considering the proportion of well success and average age of well in SMAs and NSMAs along the Uttara Pinakini river basin in Gauribidanur taluk, Karnataka

Type	Sand mining area			Non-sand mining area		
	NR/acre (₹)	Excepted NR/acre (a) (₹)	Excepted NR/acre (b) (₹)	NR/acre (₹)	Excepted NR/acre (a) (₹)	Excepted NR/acre (b) (₹)
Open well	–	–	–	10605	5090	89879
Filter point well	6635	2123	13740	14145	10327	125021
Bore well	10775	9158	82262	11684	10398	101763
Overall wells	8705	5135	41913	12143	8500	109218

The proportion of successful (average age) open wells, filter point wells and borewells are 0 (0), 0.32 (7) and 0.85 (10) in SMAs, whereas in the NSMAs, it is 0.48 (22), 0.73 (14) and 0.89 (11) for open wells, filter point wells and borewells. Considering all the wells, the proportion of successful (average age) of wells are 0.7 (15) and 0.59 (9) respectively in SMA and NSMA. Expected NR/acre (a) = NR/acre × Proportion of well success
Expected NR/acre (b) = $\sum$ (a)/$(1+i)^t$ where 'i' is the interest rate taken as 2% and 't' is average age of well

Table 4.10 Percentage difference of net returns per acre considering the proportion of well failure and average age of well along the Uttara Pinakini river basin in Gauribidanur taluk, Karnataka

Type of wells	Per cent difference of NR/acre in sand mining and non-sand mining area	Per cent difference of NR/acre (a) in sand mining and non-sand mining area	Per cent difference of NR/acre (b) in sand mining and non-sand mining area
Open wells	–	–	–
Filter point wells	53	79	89
Bore wells	7	11	19
Overall wells	28	40	61

Table 4.11 Sources of income in SMAs and NSMAs along the Uttara Pinakini river basin in Gauribidanur taluk, Karnataka. (₹ per farm per year)

Source	Net income per farm in sand mining (₹)	Share of different farm activities per rupee of income from sand mining (₹)	Net income per farm in non-sand mining (₹)	Share of different farm activities per rupee of income from agriculture (₹)
Agriculture	26298 (31)	0.53	47404 (60)	1
Livestock	7532 (9)	0.15	9054 (11)	0.19
Sand mining	49250 (59)	1	0 (0)	0
Sericulture	3500 (4)	0.07	16937 (21)	0.35
Others[a]	10575 (12)	0.21	5600 (7)	0.12
Total	82437 (100)		78995 (100)	

Figures in parentheses represent percentage to the respective total
[a] Other activities are petty shop business, milk vendor, electrical worker, contractor, tractor driver and government employee

(₹ 9158) compared with NSMAs (₹ 10,398). Further, by discounting the net returns per acre from borewells considering the average age of wells, the net returns per acre declined by 19 % in SMAs (₹ 82,262) compared with NSMAs (₹ 101,763) (Tables 4.9 and 4.10).

4.14 Sources of Income

In SMAs, about 50 % of the overall income is contributed by sand mining activity followed by agriculture (31 %), other allied activities (12 %) and livestock (9 %). In NSMAs, 61 % of the income is derived from seasonal crops (agriculture) followed by sericulture (21 %) and livestock (11 %). In SMAs, for every rupee income from sand mining activity, the farmer gets 0.53 ₹ from agriculture, 0.21 ₹ from other allied sericulture and 0.12 ₹ from livestock activities. This indicates the wide gap in income earned between sand mining activity and agriculture. This reflects the increasing rate of exploitation of sand resources in the region (Table 4.11).

Table 4.12 Net income per acre from different activities in SMAs along the Uttara Pinakini river basin in Gauribidanur taluk, Karnataka (₹ per acre per year)

Activities	Number of farmers	Agriculture (A)	Livestock (L)	Sand mining (M)	Sericulture (S)	Others (O)	Total
A+L	11 (28)	8699 (76)	2776 (24)	NA	NA	NA	11475 (100)
A+L+M	17 (42)	10241 (18)	1768 (3)	45814 (79)	NA	NA	57823 (100)
A+L+S	3 (8)	14155 (27)	7022 (13)	NA	31111 (60)	NA	52288 (100)
A+L+O	9 (22)	8965 (26)	5256 (15)	NA	NA	20142 (58)	34363 (100)
Total	40 (100)						

Figures in parentheses represent percentage to the respective total. *NA* Not applicable
A+L agriculture+livestock, *A+L+M* agriculture+livestock+sand mining
A+L+S agriculture+livestock+eericulture, *A+L+O* agriculture+livestock+other activities

4.15 Net Income per Acre from Different Activities

In SMA, 42 % of sample farmers were engaged in agriculture + livestock + sand mining and the rest in agriculture + livestock (28 %), agriculture + livestock + sericulture (8 %) and agriculture + livestock + other activities (8 %). The net return per acre was the highest (₹ 57,823) from agriculture + livestock + mining activities compared with other activities, of which 79 % was from sand mining activity. Compared to agriculture, sand mining is remunerative since there is no initial investment and no risk of crop failure due to disease or monsoon failure; instead, there is an assured income provider due to an assured market. As sand is a natural resource, farmers mine from the riverbed and supply to sand suppliers. Thus, most of the farmers who are in the proximity of a riverbed are involved in sand mining (Table 4.12).

In NSMAs, out of 30 sample farmers, 46 % were engaged in agriculture + livestock, followed by agriculture + livestock + sericulture (37 %) and agriculture + livestock + other activities (17 %). The net return per acre is the highest from agriculture + livestock + sericulture activity (₹ 31295) compared with agriculture + livestock (₹ 13441) (Table 4.13).

4.16 Access to Groundwater Resources

Open wells are of shallow depth of around 25–30 ft near the streams and sand mining is to a depth of 20–25 ft. This has a drastic impact on riparian groundwater resources resulting in drying up of all open wells in SMAs. Among the open wells in NSMAs, for every rupee of irrigation cost, farmers are realizing a net return of ₹ 13.80 as the sand along riverbed facilitates percolation and recharge of groundwater (Table 4.14).

Table 4.13 Net income per acre from different activities in NSMAs along the Uttara Pinakini river basin in Gauribidanur taluk, Karnataka (₹ per acre per year)

Activities	Number of farmers	Agriculture (A)	Livestock (L)	Sericulture (S)	Others (O)	Total
A+L	14 (46)	11950 (89)	1490 (11)			13441 (100)
A+L+S	11 (37)	13767 (44)	3215 (10)	14312 (45)		31295 (100)
A+L+O	5 (17)	16761 (52)	2263 (11)		7466 (36)	20491 (100)
Total	30 (100)					

Figures in parentheses represent percentage to the respective total

A + *L* agriculture + livestock, *A* + *L* + *S* agriculture + livestock + sericulture

A + *L* + *O* agriculture + livestock + other activities

Table 4.14 Access to groundwater resources for irrigation in NSMAs and SMAs from *open wells* along the Uttara Pinakini river basin in Gauribidanur taluk, Karnataka

Particulars	Sand mining areas	Non-sand mining areas
Total wells	All the open wells have dried up in sand mining area	17
Number of functional wells		8
Number of failed wells		9
Proportion of well failure		0.52
Gross irrigated area (acre)		30.5
Groundwater (GW) used per well (acre-inches)		20.53
GW used per acre (acre-inches)		5.29
Total GW used		161
Irrigation cost per acre-inch of GW used (₹)		144
Net returns per well (₹)		40215
Net returns per acre (₹)		10605
Total net return (₹)		321723
Total amortized cost (₹)		47385
Amortized cost per well (₹)		3755
Amortized cost per functioning well (₹)		5919
Negative externality (₹)		2164
Net returns per acre-inch of GW used (₹)		1998
Net returns per rupee of irrigation cost (₹)		13.8

1. Negative externality = Amortized cost per functioning wells minus amortized cost per well
2. Amortized cost per well = Total amortized cost of all wells/Total number of wells
3. Amortized cost per functioning well = Total amortized cost of all wells/Total number of functioning wells
4. Proportion of well failure = Number of failed wells divided by Total number of wells
5. Net return per acre-inch of groundwater used = Total net returns/Total groundwater used
6. Net returns per rupee of irrigation cost = Net return per acre-inch of groundwater used/Irrigation cost per acre-inch of groundwater used (₹)

The filter point well is a tubular structure piercing through the sand portion of the sand bed with a blank and slotted pipe surrounded by a gravel pack. Usually the filter point wells are dug near the riverbed or within the stream. Due to sand mining, filter point wells are quickly affected. As a result, the proportion of failure of filter point wells was 152 % in SMAs (0.67) compared with NSMAs (0.27). The effect of sand mining is reflected in the declining groundwater level in riparian areas. The groundwater extracted per well was 50 % lower in SMAs (13.36 acre-in. per well per year) compared with NSMAs (27.26 acre-in. per well per year). A high rate of failure of filter point wells in SMAs resulted in increased amortized cost of irrigation per acre-inch of groundwater used by 113 % compared with NSMAs. This high cost of irrigation led to increased cost of cultivation resulting in the lowering of net return per acre by 32 % in SMA (₹ 6635 per acre) compared with NSMA (₹ 14,147 per acre).

The additional investment incurred due to failure of wells can be considered as an indicator of externality. Externality is estimated as the difference between amortized costs per functional well minus the amortized cost per well. Thus, the negative externality was 531 % higher in SMAs (₹ 7287) compared with NSMAs (₹ 1154). Due to high irrigation cost, the net returns per rupee of irrigation cost were 59 % lower in SMAs (₹ 4.67) compared with NSMAs (₹ 11.55) (Table 4.15).

The proportion of failure of borewells was 36 % higher in SMAs as compared with NSMAs. The rate of failure of borewells in sand mining (36 %) is lower compared with filter point wells (152 %) since they are drilled away from streams to a depth of 400–650 ft. As the drilling of borewells involves higher initial investment due to failure of borewells, the amortized cost of irrigation is higher by 63 % in SMAs compared with NSMAs. This is reflected in a higher irrigation cost by 131 % in SMAs (₹ 307 per acre-inch) as compared with NSMAs (₹ 109 per acre-in). By considering the irrigation cost, net returns per rupee of irrigation cost are lower by 47 % in SMAs (₹ 5 per acre) as compared with NSMAs (₹ 10.3 per acre) (Table 4.16).

By considering all the wells in SMAs and NSMAs, gross irrigated area per well was 19 % lower and groundwater extraction per acre was 44 % lower due to negative externality induced by sand mining on groundwater recharge. The proportion of well failure is 53 % higher in SMA (0.46) compared with NSMA (0.3). Groundwater extracted per well was 44 % lower in SMA (17.69 acre-in. per year) compared with NSMA (32.12 acre-in. per year). The externality cost involved in drilling of additional wells due to initial and premature failure was 215 % higher in SMA (₹ 4189 per well) compared with NSMA (₹ 1328 per well). The net return per rupee of irrigation cost was 43 % lower in SMA compared with NSMA (₹ 11.88 per acre-inch) due to higher cost of irrigation in SMAs (₹ 4.32 per acre-in.) (Table 4.17).

Table 4.15 Access to groundwater resource for irrigation in SMAs and NSMAs from *filter point well (FPWs)* along the Uttara Pinakini river basin in Gauribidanur taluk, Karnataka

Particulars	Sand mining areas (SMA)	Non-sand mining areas (NSMA)	Difference between SMA and NSMA	Per cent difference over NSMA
Total number. of FPWs	25	22	NR	NR
Number of functional wells	8	16	NR	NR
Number of failed wells	17	6	NR	NR
Proportion of well failure	0.68	0.27	0.41	152
Gross irrigated area (acres)	24	52	NR	NR
Gross irrigated area per functional well	3	3.25	0.25	7
GW extracted per well (acre-inches)	13.63	27.26	−13.6	−50
GW extracted per acre (acre-inches)	4.5	8.39	−3.89	−46
Total GW used	109.16	436	−327	−75
Irrigation cost per acre-inch of GW used (₹)	312	146	166	113
Net returns per FP well (₹)	19905	45977	−26072	−56
Net returns per acre (₹)	6635	14147	−7512	−53
Total net return (₹)	159240	735644	NR	NR
Total amortized cost (₹)	85731.2	79298	6433	8
Amortized irrigation cost per well (considering all wells) (₹)	3429	3802	−373	−10
Amortized cost per functioning well (₹)	10716	4956	5760	116
Negative externality (₹)	7287	1154	6133	531
Net return per acre-inch of groundwater used (₹)	1458	1687	−229	−14
Net returns per rupee of irrigation cost (₹)	4.67	11.55	−6.88	−59

NR not relevant

1. Negative externality = Amortized cost per functioning wells minus amortized cost per well
2. Amortized cost per well = Total amortized cost of all wells/Total number of wells
3. Amortized cost per functioning well = Total amortized cost of all wells/Total number of functioning wells
4. Proportion of well failure = Number of failed wells divided by total number of wells
5. Net return per acre-inch of groundwater used = Total net returns/Total groundwater used
6. Net returns per rupee of irrigation cost = Net return per acre-inch of groundwater used/Irrigation cost per acre-inch of groundwater used (₹)

Table 4.16 Access to groundwater resources for irrigation in SMAs and NSMAs from *borewells* along the Uttara Pinakini river basin in Gauribidanur taluk, Karnataka

Particulars	Sand mining areas (SMA)	Non-sand mining areas (NSMA)	Difference between SMA and NSMA	Per cent difference over NSMA
Total wells	38	17	NA	NA
Number of functional wells	32	15	NA	NA
Number of failed wells	6	2	NA	NA
Proportion of well failure	0.15	0.11	0.04	36
Gross irrigated area (acre)	103	69.5	NA	NA
Groundwater (GW) used per well (acre-inches)	21.76	48.59	−26.83	−55
GW used per acre (acre-inches)	6.97	10.47	−3.5	−33
Total GW used	718	728	−10	−1.37
Irrigation cost per acre-inch of GW used (₹)	307	109	198	181
Net returns per well (₹)	34682	54548	−19866	−36
Net returns per acre (₹)	10775	11684	909	8
Total net return (₹)	1109825	818223	NA	NA
Total amortized cost (₹)	221041	85172	135869	1.5
Amortized cost per well (₹)	5816	5010	806	16
Amortized cost per functioning well (₹)	6907	5678	1229	22
Negative externality (₹)	1091	668	423	63
Net returns per acre-inch of GW used (₹)	1545	1123	422	37
Net returns per rupee of irrigation cost (₹)	5.0	10.3	−5.3	−51

NR not relevant

1. Negative externality = Amortized cost per functioning wells minus amortized cost per well
2. Amortized cost per well = Total amortized cost of all wells/Total number of wells
3. Amortized cost per functioning well = Total amortized cost of all wells/Total number of functioning wells
4. Proportion of well failure = Number of failed wells divided by total number of wells
5. Net return per acre-inch of groundwater used = Total net returns/Total groundwater used
6. Net returns per rupee of irrigation cost = Net return per acre-inch of groundwater used/Irrigation cost per acre-inch of groundwater used (Rs.)

Table 4.17 Access to groundwater resources for irrigation in NSMAs and SMAs from all wells along the Uttara Pinakini river basin in Gauribidanur taluk, Karnataka

Particulars	Sand mining areas (SMA)	Non-sand mining areas (NSMA)	Difference between SMA & NSMA	Per cent difference over NSMA
Total number of wells	76	57	NR	NR
Number of functional wells	40	40	NR	NR
Number of failed wells	35	17	NR	NR
Proportion of well failure	0.46	0.3	0.16	53
Gross irrigated area (acre)	127	152	NR	NR
Gross irrigated area per well (acres)	3.1	3.8	−0.7	−18.42
GW extracted per well (acre-inches)	20.67	32.12	−11.44	−35.6
GW extracted per acre (acre-inches)	5.73	8.05	−2.3	−28
Total GW extracted	827	1325	497	37
Amortized cost of irrigation per acre-inch of GW (₹)	309	133	176	133
Net return per well (₹)	27638	46913	−19274	−41
Net return per acre (₹)	8705	12143	−3438	−28
Total net return (₹)	1105535	1875590	NR	NR
Total amortized cost of irrigation (₹)	306772	211855	NR	NR
Amortized cost per well (₹)	4622	4189	433	10
Amortized cost of irrigation per functioning well (₹)	8811	5517	3294	60
Negative externality per well(₹)	4189	1328	2861	215

Table 4.17 (continued)

Particulars	Sand mining areas (SMA)	Non-sand mining areas (NSMA)	Difference between SMA & NSMA	Per cent difference over NSMA
Net return per acre-inch of GW used (₹)	1336	1602	−265	−16
Net return per rupee of amortized cost of irrigation (₹)	4.32	11.88	−7.56	−63

Wells include open wells, filter point wells and borewells
NR Not relevant
1. Negative externality = Amortized cost per functioning wells minus amortized cost per well
2. Amortized cost per well = Total amortized cost of all wells/Total number of wells
3. Amortized cost per functioning well = Total amortized cost of all wells/Total number of functioning wells
4. Proportion of well failure = Number of failed wells divided by total number of wells
5. Net return per acre-inch of groundwater used = Total net returns/Total groundwater used
6. Net returns per rupee of irrigation cost = Net return per acre-inch of groundwater used/Irrigation cost per acre-inch of groundwater used (₹)

4.17 Economics of Irrigation Considering Distance from Stream

Considering the location of wells within a distance of 120 ft from SMA, the probability of well failure is 0.52 in SMA, while in NSMA the probability of well failure is 0.33. The net return per acre of irrigated area is higher in NSMA irrespective of the distance from the river stream, compared with SMA. The amortized cost per functioning well is higher in SMAs (₹ 10,412 per well, ₹ 13,292 per well and ₹ 8241 per acre-inch of groundwater) for wells located within 120 ft from the stream than all the wells in NSMAs (₹ 4581 per well, ₹ 5030 per well and ₹ 6141 per acre-inch of groundwater). Thus, negative externality per well is higher for all wells located within 120 ft from the stream in SMAs compared with any well irrespective of distance in NSMAs. This clearly demonstrates the effect of sand mining externality in agriculture.

The groundwater extracted per well (24.76 acre-in.) in the SMA located beyond 120 ft matched with the wells located within 30 ft in NSMA (26.45 acre-in.). This apparently demonstrates that distance from the stream in sand mining has a large negative influence on the groundwater extracted per well. The groundwater extracted from wells located within 30 ft from the stream in NSMA is at least 100 % higher compared with SMA (Table 4.18).

Table 4.18 Economics of irrigation according to distance from river stream in SMAs and NSMAs along the Uttara Pinakini river basin in Gauribidanur taluk, Karnataka

Particulars	Sand mining (n=40)			Non-sand mining (n=30)		
	Distance from streams (ft)			Distance from streams (ft)		
	0–30	30–120	120–1500	0–30	30–120	120–1500
Total number of wells	25	25	26	24	12	21
Failed wells	17	13	5	8	4	5
Functioning wells	8	11	21	16	8	16
Proportion of well failure	0.68	0.52	0.19	0.33	0.33	0.23
Net return per acre of GIA (₹)	6824	7509	10064	9732	10962	12895
GIA (acres)	26.5	44	56.5	50.5	34.5	67.5
Expected net return per acre of GIA (₹)	2183	3604	8151	6520	7344	9929
Amortized cost per well (₹)	3332	4785	6656	3187	3353	4679
Amortized cost per functioning well (₹)	10412	13292	8241	4581	5030	6141
Irrigation cost per acre of GIA (₹)	861	569	332	173	295	127
Total groundwater extracted (acre-inches) for the sample farmers	96.69	209.95	520	423.22	136.31	769
Groundwater extracted per functioning well (acre-inches)	12.08	19.08	24.76	26.45	17.04	48
Negative externality per well (₹)	7080	8507	1585	1394	1677	1462

GIA gross irrigated area

1. Excepted net return per acre of GIA = (Net return per acre of GIA × Proportion of well failure)
2. Negative externality = Amortized cost per functioning wells minus amortized cost per well
3. Amortized cost per well = Total amortized cost of all wells/Total number of wells
4. Amortized cost per functioning well = Total amortized cost of all wells/Total number of functioning wells
5. Proportion of well failure = Number of failed wells divided by total number of wells

4.18 Details of Irrigation Wells Before and After Mining in Sand Mining Areas

Sand along the riverbed serves as a source of recharge of rainwater and retains moisture in the soil, and so removal of sand has led to groundwater depletion reflected in increase in well failure. The number of functioning open wells before mining was seven and groundwater yield was 2025 GPH. But after mining, there has been a complete failure of all open wells. The functioning filter point wells and borewells before mining were 23 and 42, respectively, and the groundwater yield was 2448 and 2569 GPH, respectively. But after mining, the number of filter point

Table 4.19 Details of groundwater yield of irrigation wells before and after SMA along the Uttara Pinakini river basin

Well type	Before mining (1997) (n=40)		After mining (2003) (n=40)		Difference		Per cent difference	
	No. of wells	Well yield (GPH)	No. of wells	Well yield (GPH)	No. of wells	Well yield (GPH)	No. of wells	Well yield (GPH)
Open well	7	2025	0	0	7	2025	100	100
Filter well	23	2445	8	1348	15	897	65	36
Borewell	42	2568	33	1658	9	910	21	35

Since 15 years, sand has been mined in these areas. However, villagers felt the adverse impacts since 5–6 years
GPH gallons per hour

wells and borewells fell to 8 and 33 among the sample farmers and water yield was 1348 GPH and 1658 GPH, respectively. The water declined by 100, 36 and 35% in open wells, filter point wells and borewells, respectively, in SMA (Table 4.19).

4.19 Opportunity Cost of Providing Agricultural Land for Transporting Sand

About eight sand mining farmers, forming 20% of the sample, offered some space in their agricultural land for the trucks to convey the loaded sand from the stream to the approach road. The road length provided by these eight farmers in terms of space is estimated to be 7.75 acres of land area at the rate of about one acre per farmer. The total net income of these eight farmers is ₹ 260,000 by providing an approach road to transport sand (at the rate of ₹ 50 per load at each farmers land) and ₹ 73,617 is from agriculture. However, this income is not realized every year, since it depends on the availability of sand from the stream and the demand for sand. Upon enquiry with sand mining farmers, it is estimated that this income would accrue once in 3 years. Thus, the income realized per year is amortized for 3 years at a modest rate of 2% to get the net income from load linkage per year. In addition, once the farmer offers the road space, since around 50–60 truckloads pass through the road per day, the soil gets hardened. In addition, the additional area on either side of the road cannot be effectively put to use for agriculture purposes. Thus, even though the road space for transporting comes to around half an acre per farm, it is taken as one acre, giving a margin to these factors.

The net income per acre for road linkage land considering opportunity cost of agricultural land is estimated as (total net income from road linkage−total net income from agriculture) divided by the total net cropped area of all eight farmers. The net cropped area of eight sample farmers who form 20% of the sample farmers

Table 4.20 Opportunity cost of providing agricultural land for transporting sand

Total number of sample farmers in sand mining area	40
Number of sample farmers involved in sand mining activity	17
Number of farmers involved in sand mining who provided road linkage to transport sand from river stream	8
Total cultivated area lost due to road linkage for linking sand transport from the river stream to approach road (acres)	7.75
Net income earned by providing road linkage of 7.75 acres for transporting sand at the rate of ₹ 50 per load per truck X three trucks per day on half acre (load linkage space) for 360 days for 8 farmers (₹)	260,000
Net income per acre by providing road linkage to transport sand (₹) (260,000/7.75)	33,548
Estimated net income from 7.75 acres of agricultural land (₹)	77,438
Net income per acre of agricultural land (₹)	9992
Total net cropped area of the eight sample farmers who provided road linkage to transport sand (acres)	27.5
Net income per acre of road linkage land considering the opportunity cost of agricultural land (₹) (260000−33548)/27.5=6637	*6639
Amortized net income per acre of road linkage land considering the opportunity cost of agricultural land for 3 years at 2%	2581

in the SMA is taken as a denominator since these sand miners are basically farmers, for whom sand mining is just a subsidiary and not a main activity. All sand miners do not keep land for road linkage. Only those sand miners located closer to the approach road as well as whose land serves as an effective link between the approach road and stream will devote a part of their land for linking the stream with the approach road. The net return per acre of road linkage considering the opportunity cost of agricultural land is ₹ 6637 and amortized net return per acre is ₹ 2581. (Table 4.20)

4.20 Analysis of Sand Mining Accounting for Externalities

Sand mining has a direct impact on groundwater recharge in the riparian areas. The difference in amortized cost of irrigation per acre between sand mining and non-sand mining gives the additional cost incurred by farmers in SMA. This is estimated to be ₹ 1624. In addition, there will be a decrease in returns due to sand mining.

Since the sand from the riverbed has to be transported to the main road across agricultural fields, sand miners will have to pay for the road linkage space. In addition, farmers get an opportunity once in 3 years to provide road linkages, and hence it is amortized for 3 years at 2%. The difference between net return per acre in NSMA and SMA is an externality cost due to sand mining. The total externality

Table 4.21 Partial budgeting analysis of sand mining by farmers without considering externalities due to sand mining in Gauribidanur taluk, Karnataka

Cost (A)	Cost (B)
Increase in cost Nil	Decrease in cost Nil
Decrease in returns 1. Opportunity return from cultivated land forgone for providing road linkage to transport the sand per acre = ₹ 2659 2. Net return per acre from agriculture in non-sand mining area minus net return per acre from agriculture in sand mining area excluding groundwater irrigation cost per acre = (₹ 13,680 − ₹ 10,227) = ₹ 3453 A = Total = ₹ 2659 + ₹ 3453 = ₹ 6112	Increase in returns 1. Annual net returns per acre from sand mining activity for the farmer = ₹ 12,362 (= total income from sand extraction for all farmers divided by gross irrigated area B = Total = ₹ 12,362
Net returns in sand mining area = B-A = ₹ 12,362 − 6112 = ₹ 6250	

Explanation for item 1 and 3:

1. Opportunity cost of cultivated land foregone for road linkage is estimated as below:
   a. Net return from agriculture per acre in sand mining area excluding the irrigation cost = ₹ 9499
   b. Number of sample farmers who provided road space = 8
   c. Total cultivated area of eight sample farmers who provided road linkage in SMA = 27.5 acres
   d. Estimated cultivated area lost due to provision of road space for transporting sand by sample farm = 7.75 acres
   e. Road linkage space/total cultivated land area = 7.75/27.5 = 0.28 acres
   f. Opportunity cost of cultivated land for providing road link space = (a × e) = 9499 × 0.28 = ₹ 2659
2. Annual net returns incurred by sand mining by sample farmers = ₹ 1,909,929.

Annual net returns per acre from sand mining activity for the farmer = 1,909,929/154.5 = 12,362

cost amounts to ₹ 711 per acre and returns to sand sales are assumed to accrue once in 3 years. The returns are amortized at2 % interest rate (Tables 4.21 and 4.22).

Results indicate that when externality is considered, the net return in SMA is minus ₹ 771 per acre, whereas without considering externalities the net returns account to ₹ 6250. This apparently indicates that farmers are overextracting sand and groundwater, as a result enjoying a pseudo-illusionary return of ₹ 6250 per acre, while they are in fact losing ₹ 711 per acre due to the sand mining externality (Tables 4.21 and 4.22).

4.21 Economics of Sand Transportation from Uttara Pinakini Riverbed in Gauribidanur Taluk to Bangalore Megacity

The demand function of sand was estimated in Bangalore using data from 30 sand suppliers. On an average, a sand truck or lorry supplied 25 loads per month realizing a gross return of ₹ 4500 per load and net return of ₹ 1285 per load. The sand loader

Table 4.22 Partial budgeting analysis of sand mining in the farm accounting for externalities along the Uttara Pinakini river in Gauribidanur taluk of Karnataka

Cost (A)	Cost (B)
Increase in cost due to sand mining activity in the farm through groundwater irrigation *Externality cost* 1. Amortized cost of ground water irrigation per acre in sand mining area minus amortized cost of groundwater irrigation in non-sand mining area =(₹ 3008−1384)=₹ 1624	Decrease in cost due to sand mining activity on agriculture 4. Since 7.75 acres out of 27.75 acres are lost for providing road linkage, on an average 0.28 acre is lost. Thus the decrease in cost of cultivation is (0.28 × ₹ 3500 per acre)—₹ 980, is the savings in cost of cultivation due to the land foregone for transporting sand
Decrease in returns due to sand mining activity in the farm *Externality cost in terms of returns foregone due to provision of road space for transporting sand, linking the sand source and the main road* 2. Amortized opportunity return for cultivated land forgone for providing road linkage to transport sand per acre=₹ 924 3. Net return per acre from agriculture in non-sand mining area minus net return per acre from agriculture in sand mining area in the farm excluding groundwater irrigation cost=(₹ 13, 680−10,227)=₹ 3453 A=Total=₹ 1624+924+3453=₹ 6001	Increase in net returns due to sand mining activity on the farm 5. Amortized net returns per acre from sand mining=₹ 4310 B=Total=₹ 980+4310=5290
Net returns in sand mining areas=B-A=₹ 5290−6001=−₹ 711	

Assumptions of partial budgeting analysis

1. Riparian farmers undertake sand mining, as they are closer to river stream carrying sand and due to squatter's right
2. Though 17 farmers out of the sample of 40 farmers in sand mining area are involved in sand mining activities gross cropped area of all sample farmers is considered in estimating externality due to sand mining
3. Returns from road linkage, opportunity cost of labour and net return per acre from sand mining are amortized for 3 years at an interest rate of 2 %, since sand is assumed to accumulate once in 3 years due to pattern of rainfall
4. Amortized cost of irrigation per acre = Amortized cost of investment on all wells, conveyance, pump set and pump house

Explanation for item 2 and 5:

1. Opportunity cost of cultivated land foregone for road linkage is estimated as below
   a. Net return from agriculture per acre in sand mining area excluding the irrigation cost = ₹ 9499
   b. Number of sample farmers who provided road space = 8
   c. Total cultivated area of eight sample farmers who provided road linkage in sand mining area = 27.5 acres
   d. Estimated cultivated area lost due to provision of road space for transporting sand by sample farm = 7.75 acres
   e. Road linkage space/total cultivated land area = 7.75/27.5 = 0.28 acres
   f. Opportunity cost of cultivated land for providing road link space =(a X e)=9499 X 0.28=₹ 2659
   h. Externality cost for providing road links is amortized for 3 years @ 2 %=₹ 924

The opportunity cost for road link space of ₹ 2021 is realized approximately once in 3 years as sand mining activity is assumed to take place once in 3 years due to pattern of rainfall

2. Annual net returns incurred by sand mining by sample farmers = ₹ 1909929
   a. Annual net returns per acre from sand mining activity for the farmer = 1909929/154.5 = 12362
   b. The annual net returns per acre from sand mining is realized approximately once in 3 years as sand mining activity is assumed to take place once in 3 year due to pattern of rainfall
   c. The amortized net return per acre at the rate of 2 %=₹ 4310

Table 4.23 Economics of sand transportation from Uttara Pinakini riverbed in Gauribidanur taluk to Bangalore city

Cost per truck of sand (2003)	Return per truck of sand (2003)
1. Cost of mining sand from stream and loading to truck =₹ 1000 2. Royalty charges paid to government per truck-load =₹ 45 3. Fuel cost incurred for transporting sand from Gauribidanur to Bangalore city =₹ 1200 4. Cost of unloading from lorry at delivery point =₹ 300 5. Driver charges for transporting per truckload =₹ 200 6. Depreciation (@ ₹ 333 per load + Estimated Repairs @ ₹ 117 = ₹ 450 7. Miscellaneous expenses =₹ 120 Total =₹ 3365	Price per truckload of sand at Bangalore city =₹ 4200
Net income = Total returns—Total cost =₹ 4200 − 3365 = ₹ 835	

incurred a cost of ₹ 2915 per load at the place of sand mining which includes ₹ 1000 per load towards cost of sand mining from stream and loading to lorry, ₹ 1200 towards fuel cost incurred for transportation of sand from the source to Bangalore, ₹ 45 per load towards payment of royalty, ₹ 300 per load as payment towards loading and unloading from lorry, ₹ 200 incurred for driver charges per load, depreciation cost ₹ 450 per truckload and ₹ 120 incurred for other miscellaneous expenses. The sand miner sells the sand at ₹ 4200 per load. Net returns of ₹ 835 have been earned by sand transporters. Usually, the sand miner waits for 1–2 day/s for the sand to be disposed off at the construction site where sand is used for civil works (Table 4.23).

4.22 Cobb Douglas Function for Demand for Sand in Megacities

The demand function for sand for Bangalore megacity was estimated in order to estimate the price elasticity of demand for sand. The dependent variable was the logarithm of the volume of sand demanded per month per truck. The explanatory variables were the logarithm of the distance transported per truck each time in kms, the logarithm of the number of days taken for sale at Bangalore megacity and the logarithm of price of sand per cft (Table 4.24).

The estimated demand function for sand is $Y = 715\ (\text{Distance})^{0.323}$ $(\text{No. of sale days})^{0.0036}$ $(\text{Sand Price})^{-0.876}$, where Y = volume of sand transported or sold in cubic feet, D = distance in kilometres, no. of sale days is the number of days taken for sand to be sold in the megacity, and sand price is price per cubic feet of sand.

Sand demanded per day has a price elasticity of −0.88. This implies that for a 1 % increase in price of sand per cubic foot, the quantity demanded of sand reduces

Table 4.24 Estimation of demand function for sand (volume in cubic feet) in the megacity of Bangalore, 2003 (Cobb Douglas function)

Dependent variable: Natural log of volume of sand demanded per month per lorry (Geometric mean of sand demanded per month = 9897 cft)

Explanatory variables	Regression coefficient	t-value	R^2 = 0.873	Geometric mean
Ln of Intercept	6.572	22.712		9897 cft
Ln of distance travelled per truck per time (Kms)	0.323[a]	6.473		106.06
Ln of sale days (days)	0.003622	0.237		1.785
Ln of price per cft (₹)	−0.876[a]	−11.629		10.45

[a] Significant at 1 %

by 0.88 %. The elasticity of sand demanded with respect to distance is 0.32 and elasticity of sand demanded with respect to number of days taken to sell the sand in Bangalore is 0.0036. Thus, with increase in distance, sand being indispensable, the demand also increases; even though the demand is inelastic with respect to distance. However, the price or demand for sand is almost unrelated with the number of days taken to sell the sand in Bangalore due to the modest coefficient of 0.0036. This shows that the number of days taken for disposal of sand at the point of destination does not influence the price of sand.

Thus, the price elasticity of demand is negative and inelastic implying that with a 1 % increase in sand price per cubic feet from the mean price of ₹ 10.45 per cubic foot, the quantity demanded for sand reduces by 0.88 %. For a 1 % increase in the distance from the sand source to Bangalore from the mean distance of 106 Km, the demand for sand increases by 0.32 %. Since there is no substitute for sand and sand with desirable quality for civil works is available only in the river stream, the price/demand for sand with distance is inelastic.

On an average, one truckload has 350 cft of sand. If the price of sand per cubic feet increases by 1 ₹ from the mean price of ₹ 10.45 per cubic feet, the quantity demanded of sand reduces by 35.36 ft^3. Similarly if the distance of sand transportation from Bangalore to the sand mining source increases by one kilometre above the mean distance of 106 Km, the demand for sand increases by 12.86 ft^3.

Documentation of Travails Faced by Farmers and Stakeholders due to Sand Mining Groundwater is the major source of irrigation in Gauribidanur taluk receiving scanty rainfall of 679 mm. Due to sand mining, there is depletion in groundwater levels affecting riparian farmers. In addition to groundwater depletion, the problems faced by farmers and stakeholders are vegetation loss on shoulders of the riverbed, undermining of roads and rail bridges. They also complained of noise and dust pollution during excavation, loading operations of and in the sand mining area (Table 4.25).

Table 4.25 Documentation of travails due to sand mining activity along the Uttara Pinakini river basin in Gauribidanur taluk, Karnataka

Problems due to sand mining	% of farmers
Groundwater depletion	95
Shift in cropping pattern from commercial crops to subsistence crops	78
Scarcity of agricultural labourers	72
Vegetation loss on either side of riverbed	87
Affected groundwater conveyance pipes from filter point wells to farm fields	80
Creation of pits in riverbeds disrupt natural flow of water	49
Undermining of road and rail bridges are destabilizing the bridges	52
Conflict between villagers and mining labourers creating tension disturbing the social life	56
Increased use of unplanned access roads has lead to soil compaction and vegetation damage, collateral damage to roads due to increased number of sand transporting lorries	65
Intrusion of hard water into soft water in the wells	45
Addiction to alcohol	30
Dust allergy	62
Noise and dust pollution	60
Damage to check dam	40
Low prices to cocoon due to dust accumulation on mulberry leaves	38
Soil erosion	20

It is heartening to note that a group of stakeholders led by an engineer, D.A. Chowdappa, environmentalist of Doddakugodu village, Gauribidanut taluk, filed a public interest litigation in the high court of Karnataka (Writ Petition No.359646 of 2001) challenging the sand mining activity by miners under Article 226 and 227 constitution of India. They requested the court to grant an interim stay against sand mining activity in Gauribidanur area along the Uttara Pinakini river. They challenged the action of Department of Mines and Geology, Government of Karnataka who have accorded sand mining leases to 15 quarrying and transporting miners, stating that such intense mining is ecologically and environmentally degrading the groundwater position in the entire taluk. They stated that miners are illegally quarrying and transporting sand much against the research report by the Department of Mines and Geology, Government of Karnataka, brought out in 1986. In addition, petitioners stated that the miners engage tractors of farmers for transporting sand. This is illegal under the Karnataka Motor Vehicles Act, which permits farmers to use tractors only for agricultural purposes, for which they get a tax concession. Thus, they alleged that the Motor Vehicles Department and the police are clandestinely permitting sand mining and transporting which is not only affecting the groundwater aquifer, but also causing air and noise pollution and disturbing peace and tranquillity in villages. The honourable high court is yet to give verdict (Figs. 4.2, 4.3, 4.4, 4.5, 4.6, 4.7, 4.8, 4.9).

Fig. 4.2 Farmers engaged in sand mining activity along the Uttara Pinakini river stream in Gowribidanur taluk, Karnataka, India, 2003, Pic: MG Chandrakanth and AC Hemalatha

Fig. 4.3 Even graveyards are not spared from sand mining activity in Gowribidanur taluk, Kolar district, Karnataka, India (by MG Chandrakanth and AC Hemalatha)

Sand mining Changing the Crop Pattern from Water Intensive to Low-Water Crops in Tharati The predicament of sand mining in virtually changing the crop pattern in Tharati, eastern dry zone is highlighted. The dynamics of crop pattern changes

Fig. 4.4 Intense sand mining activity to a depth of 30 ft in Uttara Pinakini river stream in Gowribidanur taluk Sand mining, Kolar district, Karnataka, India (by MG Chandrakanth and AC Hemalatha)

Fig. 4.5 Sand mining to a depth of 30 ft in Uttara Pinakini river stream in Gowribidnaur taluk (by MG Chandrakanth and AC Hemalatha), Kalludi in Gauribidanur Taluk, Karnataka, 2003

Fig. 4.6 Sand being washed using compressor pump to remove red sting as white coloured sand fetches high price, Gauribidanur taluk, Karnataka (by MG Chandrakanth and AC Hemalatha)

Fig. 4.7 Sand being mined in agricultural field near the Uttara Pinakini river stream in Heribidanur village, Gauribidanur taluk, Karnataka, 2003 (by MG Chandrakanth and AC Hemalatha)

Fig. 4.8 Railway bridge foundation frequently being repaired due to sand mining across Uttara Pinakini river in Gauribidanur taluk, Karnataka (by MG Chandrakanth and AC Hemalatha)

Fig. 4.9 Public protesting against sand mining activity in Gauridanur town, 2002. (Source: Chowdappa, D A (2001): 'Memorandum of Writ Petition under Article 226 and 227 of the Constitution of India in the High Court of Karnataka at Bangalore, Writ petition No. 35946)

in Tumkur district in Tharati is examined considering the crop pattern followed by farmers for over a decade. The initial study by Lokesh (1998)[2] recorded the crop pattern in Tharati during 1998. Matching this with the International Crops Research Institute for the Semi-Arid-Tropics (ICRISAT) voluntary data sharing agreement (VDSA), apparently, the crops grown during 2011 were orthogonal to those in 1998. In 1998, Tharati had ample groundwater resources, as the low lying areas were surrounded by hillocks, and with good recharge, the shallow dug/open wells were providing groundwater for irrigation. With impressive groundwater resources, Tharati was the only village (in India and the world) cultivating the most water-intensive crop of *Acorus calamus* (Baje in Kannada) (sweet flag), a medicinal rhizome, which grows in swamps for 10–12 months. Considering the crop pattern on farms, 46 % of the area was irrigated, of which the major chunk of groundwater resource was for water crops such as sweet flag which occupied 56 % of the irrigated area, followed by paddy (15 %).

Drastic Reduction in Area Under Sweet Flag The area under sweet flag which was around 21 acres in 1998, reduced to 0.5 acre in 2011, a reduction of 98 %. Similarly, the area under paddy which was around 39 acres in 1998, reduced to 20 acres in 2011, a reduction of around 50 %. Even the area under groundnut crop was reduced by 90 %. However, the area under chrysanthemum and china aster, which was virtually nonexistent in 1998, is around 11 acres in 2011.

Considering the transitional probability (Table 4.1), the probability of the area under sweet flag shifting to flower crops is 0.90. Thus, the probability of staying in the present state of sweet flag is zero, while the probability of moving to the state of flower crops is 0.90. Similarly, the probability of paddy and ragi staying in the state of paddy and ragi is 0.75, while that of moving to perennial crops is 0.25. The probability of moving from the state of groundnut crops to that of paddy and ragi is 0.76, while that of moving to flower crops is 0.15. Thus, diversification holds the key for development in the post-green revolution period while specialization held the key during the green revolution period (Table 4.26).

Volte Faces—Increase in Proportion of Irrigated Area due to Reduced Groundwater! During 2011, the proportion of area irrigated in Tharati increased to 65 %, not due to increased volume of groundwater available, but due to drastic reduction in groundwater volume, and the advent of drip irrigation and the associated crop patterns. However, the area under the most water intensive crop—sweet flag, virtually reduced to zero. Thus, the Tharati crop pattern had a *volte face* treatment, with the water intensive sweet flag almost replaced by low water flower crops such as chrysanthemum and buttons. The major factors responsible for the predicament are (1) illegal sand mining which began during 2000 and is continuing unabated and (2) deforestation of hillocks. One of the reasons for excellent recharge of groundwater in dug wells during 1998, was the sumptuous sand layer which held the rain water and later percolated as groundwater in dug wells. However, due to illegal sand mining, the groundwater holding capacity of the aquifer reduced gradually. The

[2] Lokesh 1998.

Table 4.26 Transitional probabilities of cropping pattern in Tharati village in Tumkur district of Karnataka (1998–2011). (Source: Basavaraj 2014)

Crops	Paddy and Ragi	Groundnut	Chrysanthemum and China aster	Sweet flag	Perennial crops	Area in base year (1998) (Acre)	Area in Terminal year (2011) (Acre)	% change
Paddy and Ragi	0.7471	0.0000	0.0000	0.0000	0.2529	38.55	19.928	−48.31
Groundnut	0.7629	0.0000	0.1477	0.0894	0.0000	10.25	1.1	−89.27
Chrysanthemum and China aster	1.0000	0.0000	0.0000	0.0000	0.0000	0.00	11.125	–
Sweet flag	0.0000	0.0964	0.9036	0.0000	0.0000	20.95	0.5	−97.61
Perennial crops	0.0000	0.0000	0.0000	0.0000	1.0000	7.93	9.61	21.19

Perennials crops include Arecanut, Banana

denudation of forests in surrounding hillocks exacerbated the predicament. Currently, Tharati is in the conundrum of groundwater scarcity due to both manmade and climate change effects. This is further reflected in the performance of irrigation wells. During 2011, with 190 wells, 50 % were borewells and the rest, open wells. Among the borewells, only 50 % were functioning, while among dug wells only 9 % were functioning. This shows the rapid fall in the groundwater potential as reflected in the high rate of well failure of 91 % in the case of open/dug wells and 50 % in the case of borewells. Accordingly, this fall in the groundwater endowment and the corresponding change in the crop pattern by the farmers in an attempt to remain on the original isorevenue curve, have resulted in coping mechanisms such as drip irrigation and choice of low water high value crops such as chrysanthemum, buttons, virtually abandoning the *Acorus calamus*.

Sand Mining Activity in Tharati Tharati derived from Turu Hatti, meaning the habitat for cows—calves. Thus, this place had a good water source for rearing cows and calves using the tank water from Tharati tank. The Tharati village is surrounded by hillocks and the rains bring along sand due to erosion over time. The sand is deposited in the fields around the irrigation tank. Due to the sand layers, the rainwater received seeps slowly in the sand and groundwater is recharged with lateral and vertical movement to the dug wells and filter point wells. Due to this phenomenon, Tharati had sumptuous water and farmers could cultivate only *Acorus calamus* as that was the only crop which could be cultivated, since in low lying areas with sumptuous water even paddy could not be cultivated, as water was available all year round, due to the sand acting as a spongy layer. The open wells/dug wells had sumptuous water due to the sand layer which acted as a spongy layer holding rain water and enriching the open wells/dug wells. Farmers used to pump water for cultivating sweet flag which required 10–12 months standing water. There was so much groundwater that farmers had no alternative other than to cultivate sweet flag in Tharati. Farmers had open/dug wells during 1998–2000, from which they used to pump water for cultivating sweet flag which required 12 months standing water. Thus, Tharati was the only village in the world cultivating *Acorus calamus*, a medicinal plant which is currently used in ayurvedic medicines for both humans and for veterinary medicine.

Sand mining in Tharati, which commenced around 2002, seriously began hampering the groundwater recharge. The open/dug wells which were yielding water had gradually gone dry (by 2013). There was no borewell in Tharati during 2000 and there were filter point wells which were hardly 30–40 ft deep. These wells had so much water that same volume they were pumping was recharged/recouped in 24 h in the dug wells/filter point wells. Hence, sweet flag which required water to the tune of around 100 acre-in. (per acre), was supplied by these open wells/filter point wells during 2000. The price of sweet flag during 1998 was ₹ 1250 per quintal. However in 2013, in just 15 years, sweet flag, cultivated in the only village in the world using fresh water, has virtually vanished and in its place farmers are now cultivating chrysanthemum crop, due to reduced water availability, all due to sand mining activity. The price of sweet flag which was ₹ 1250 per quintal in 1998 has now shot up to ₹ 23,000 per quintal. This in itself is a *prima facie* indicator of the short supply which resulted in price rise.

The open wells/dug wells of small farmers had sumptuous water due to a sand layer which acted as a spongy layer holding rainwater and enriching the open wells/dug wells. Gradually, with the removal of sand, the dug wells failed. At present, out of 40 dug wells which were functioning during 1998–2000, no dug well is functioning. As all the dug wells and even filter point wells failed to yield water, farmers have drilled deep borewells beyond 1000 ft. Even deep borewells have now failed to yield groundwater.

Chrysanthemum is cultivated as a 7 month crop coping with reduced groundwater. It takes 4–5 months for cultivation and yields flowers for 2–3 months depending upon the availability of water, which prolongs the period of yielding flowers. Flowers are harvested every week. Farmers find it comfortable to cultivate this crop on staggered plots of 1/4 acre to 4 plots or one acre with 4 staggered plantings. Most farmers cannot afford to have more than one plot due to water scarcity also coupled with labour scarcity in management of the crop. A majority of the farmers are cultivating on 0.1 acre, earning a net return of around ₹ 30,000. The crop needs irrigation once in 4 days. As constructing/drilling a well is prohibitive costing at least ₹ 2.5 lakhs, a few farmers are cultivating chrysanthemum, purchasing groundwater paying 1/3rd of the value of the flower crop as water rental. Water markets are also vanishing since there are only a handful of farmers selling water as they have no surplus water to sell.

Sand Extraction The sand mining which is rampant in Tharati village is responsible for the groundwater predicament. Every day, it is estimated that around 20 tractor loads of tank silt and 10 truckloads of tank silt are excavated from the Tharati tank. Despite several efforts by the administration, sand mining is continuing unabated adding to the groundwater depletion. Due to the groundwater shortage, around 150 borewells are drilled and at present only 10 % of them are functioning. A 900 ft borewell drilled in 2009, cost ₹ 1.25 lakhs as drilling charges, ₹ 75,000 for the transformer to support 20 HP pump valued ₹ 70,000 and a pipe line to transport water worth ₹ 2 lakhs, totalling ₹ 470,000. In January 2013, the farmer drilled another well to a depth of 1300 ft paying ₹ 225,000 as drilling charges, but the borewell failed to yield any water. The farmer ended up spending ₹ 725,000 on both wells. Thus, well investment instead of being treated as fixed cost, becomes a variable cost. Even at 0 % interest rate, the farmer would spend around ₹ 140,000 per year, with an assumption that the existing well serves for 5 years. Assuming that this well draws around 100 acre-inches, the cost of groundwater per acre-inch =₹ 1400. Thus, the groundwater situation is alarming in the village.

Unless the illegal sand mining activity in Tharati is checked, farmers will continue to face the predicament of acute water scarcity. With the rapid reduction in the area under sweet flag (Baje in vernacular) (*Acorus calamus*), the market price of *Acorus calamus* which was ₹ 1250 per quintal during 1998 has shot up to a whopping ₹ 23,000 per quintal in 2013. The transitional probability analysis (or Markov chain analysis) indicated in Table 4.1 has evidence of this phenomenon of shift in crop pattern in Tharati (Figs. 4.10, 4.11, 4.12, 4.13, 4.14).

Income from Sand Mining In order to load one truck of sand in Tharati village in Tumkur district, labourers are paid ₹ 3000 as wages (for five labourers), and thus one farmer earns ₹ 600 as wage per truckload. Usually labourers load two truckloads of sand per day, which means they earn easily ₹ 1200 as wage per day by loading sand on to a truck. If a labourer can earn a wage of Rs 1200 per day, obviously he will ignore agriculture and becomes myopic as he may not appreciate the long-term effect of sand mining on groundwater on farms. Sand mining is the greatest culprit which has virtually depleted groundwater in many villages of Kolar, Tumkur, Mandya—Shimsha belt. Sand mining is ruining agriculture in these areas. In order to cope with the predicament, farmers shifted to cultivating chrysanthemum, which fetches them good income.

Check on House Construction/Apartment Buildings Is Required There should be limits to the growth of a city/town/metropolis. Without limits, sand mining will continue unabated at the cost of agriculture and irrigation. Urbanites have no limit on the size of their houses. The construction of oversized buildings/houses is unnecessary as the family size is growing smaller and smaller; still, highly literate people are building houses which are 30 or 45 square feet. This demands a great volume of sand, which is unfortunately mined from villages. Sand mining in the villages has already led to groundwater depletion and farmers are lured by sand truckers who pay high wages.

Fig. 4.10 Tharati groundwater scenario

Corporate Social Responsibility The software/hardware sector is largely responsible for the imbalanced growth of Bangalore metropolis, claims CSR. However, no single multinational or national firm involved in software/hardware/any other sector in Bangalore has a program to check on high rise buildings/apartments/constructions, including the Bangalore Development Authority. The construction activity has triggered large scale sand mining. The growth of Bangalore has been largely at the cost of the development of villages around Bangalore from where sand has been mined for the unsystematic growth of Bangalore.

Fig. 4.11 Sweet flag crop of pre 2000 due to abundant water. (Picture by MG Chandrakanth and GB Lokesh)

Fig. 4.12 Arecanut replacing sweet flag in 2013 due to groundwater scarcity in Tharati. (Picture by MG Chandrakanth)

Fig. 4.13 Lone farm with sweet flag in 2013 in Tharati. (Picture by MG Chandrakanth)

Fig. 4.14 Sweet flag replaced by Chrysanthemum in 2013. (Picture by MG Chandrakanth)

4.23 Conclusions

The Bangalore Development Authority, Urban Development Department, town planning and other associated authorities move in only one direction—towards urban growth. Their planning lacks, what would be the impact of this growth on the villages around Bangalore. The growth pattern of Bangalore is resulting in islands of poverty around the city. The citizens also do not realize that 'poverty anywhere is a threat to prosperity everywhere'.

The Department of Mines and Geology undertakes research in the field of groundwater in different geological conditions. However, the Department has no teeth in checking illegal sand mining. In addition, the department is not equipped to create awareness among farmers in the villages regarding the ill effects of sand mining. Even the Department of Agriculture/Horticulture, Rural Development and Panchayat Raj (RDPR) department, have no awareness programs in capacity building regarding sand mining. There are no radio programs/TV programs which air reflecting the externalities due to sand mining. Unless these departments, including the Home Department, take the issue of sand mining seriously, no single department such as Mines and Geology alone can address the issue of sand mining in any state in India. Mere legal solutions will not work, unless farmers are educated and village panchayats are empowered to take decision on sand mining. Village panchayats must be empowered to reject/not to issue licenses for sand transport from their villages. Village panchayats need support from the Home Department to address and tackle issues of sand mining and other activities leading to natural resource depletion in villages. Unless the authorities take a stern and bold step, sand mining will continue unabated, destroying agriculture, and will widen the existing income gaps resulting in non-inclusive growth.

References

Hemalatha AC (2003) Estimation of Externalities due to sand mining in water streams in affected Riparian area of Karnataka, Unpublished MSc (Agri) thesis submitted to the Department of Agricultural Economics, University of Agricultural Sciences Bangalore

Lokesh GB (1998) A resource economics study of sweet flag (Acorus calamus) in Tumkur district—Karnataka, Unpublished M. Sc (Agri) Thesis, submitted to Department of Agricultural Economics, University of Agricultural Sciences, Bangalore

Basavaraj Jamakhandi (2014) Impact of sand mining in Tharati, presentation to Department of Agricultural Economics, University of Agricultural Sciences, Bangalore 560065, April 20 2014

Chapter 5
Relationship Between Rainfall and Recharge

Here is a modest attempt to make farmers and policymakers understand and appreciate how important our groundwater resource is, especially of interest for farmers/consumers in hard rock areas of India, which form 65 % of India's area. In Karnataka, hard rock forms 99 % of the state area. What is important for the farmers is to note that the recharge to groundwater from rainfall is hardly 5–19 % in alluvial areas and much below this proportion in hard rock areas. Specifically this means that if we receive a rainfall of 750 mm, at the most, only 75 mm is the natural recharge, and the rest drains away. However, the evapotranspiration and a host of other factors are too technical and are not incorporated in this naïve understanding. To that extent, this note suffers. Here is the basic arithmetic of the complexity of groundwater recharge.

One acre of land is 4000 m^2. In a year, this land receives a rainfall of around 750 mm, which is roughly the rainfall in the hard rock areas of India. Let us now find how much of this rainfall actually contributes to groundwater in relation to how much we pump for our irrigation through the following steps:

Step 1: Convert 750 mm to meter using 750 mm = 75 cm = 0.75 m.

Step 2: Multiply the rainfall of 0.75 m with 4000 m^2 (or one acre of land) of land which receives that rainfall = 0.75 m × 4000 m^2 = 3000 m^3.

Step 3: 1 m^3 is = 1000 Litres of water. Therefore, 3000 m^3 = 3,000,000 L of water.

Step 4: 1 gallon of water = 4.54 L. Therefore 3,000,000 L of water = 660,793 gallons.

Step 5: Therefore, 660,793/22,611 = 29 ac-in (acre inches or ha cms). given 1 ac-in. = 22,611 gallons of water.

Step 6: Thus, 1 ac with 750 mm of rainfall will receive 29 ac-in. of rainwater. But only 10 % enters as groundwater = 2.9 ac-in.

Step 7: One irrigation well roughly extracts around 100 ac-in. of water in a year. This irrigation well may be serving say 3 acres on an average. Thus, for 3 ac, the total recharge is say 2.9 ac-in. × 3 ac = 8.7 ac-in. or approximately 9 ac-in. Thus, by contributing hardly 9 ac-in. of water, towards recharge, the farmer is in fact pumping out 100 ac-in. of water per well on an average every year, leaving a

M.G. Chandrakanth, *Water Resource Economics*, DOI 10.1007/978-81-322-2479-2_5

gap of around 91 or say approximately 90 ac-in. to be recharged by other areas consciously or inadvertently.

Here are the aspects the users of groundwater/farmers need to be educated.

Step 1: Usually a farmer withdrawing around 100 ac-in. of groundwater from one irrigation well is in fact getting groundwater from at least another 30 ac of land since each acre recharges only around 3 ac-in. of water per year.

Step 2: Unless groundwater recharge programs are undertaken on a massive scale, groundwater wells will face premature/initial failures. And, as these cannot be borne by small and marginal farmers, they continue to suffer due to lack of access to groundwater. Thus, we need to educate and encourage the farmers to have a groundwater recharge structure for his/her borewell on the farm by proper training and giving incentives. In addition, we can give benefits from different governmental programmes to only such farmers who adopt groundwater recharge methods as given by experts in agricultural engineering.

Step 3: Drawing lessons from the studies in the Department of Agricultural Economics, UAS Bangalore funded by the Ford Foundation, indicate that (1) the probability of well failure in the eastern dry zone is at least 0.4 (Nagaraj et al. 1994); (2) open wells/dug wells have virtually vanished due to the advent of deeper borewells over time (Chandrakanth and Romm 1990); (3) the amortized cost of groundwater ranges from ₹ 300 to 400 per ac-in (Chandrakanth and Nagaraja 2014)., considering the negative externalities due to cumulative interference among irrigation wells (Diwakara and Chandrakanth 2007), but without considering the electricity cost; (4) to lift 1 ac-in. of groundwater, the electricity cost (Chandrakanth 2009) is around ₹ 100, which implies ₹ 10,000 per year per well for 100 ac-in. and (5) thus the total cost of groundwater inclusive of electricity ranges from ₹ 400 to 500 (Chandrakanth 2009, p. 23.).

Step 4: What are the implications? (1) Our farmers and our government, if they are really concerned about the precious and scarce groundwater, should first think whether it is ethical for them to cultivate paddy/sugarcane using groundwater? (2) If they still decide to cultivate, they should realize that they are in fact using (40 ac-in. × ₹ 500 per ac-in. =) ₹ 20,000 worth of groundwater for every acre of paddy they cultivate and the double of that = ₹ 40,000 worth of groundwater for every acre of sugarcane they cultivate. These costs may vary for other parts of Karnataka, but still they may not vary substantially since 99 % of Karnataka's aquifers are hard rock aquifers; the situation is alarming. Thus, instead of cultivating paddy and sugarcane using groundwater, they should think of low-water crops which are even more economically viable than these two high-water but low-profit crops. If they still wish to pursue cultivation, they have to use micro irrigation so that water saving is realized.

Step 5: What is the solution? Farmers should realize that they are drawing a precious resource, more than gold (since gold cannot be consumed), and hence use the groundwater in such a way that they make a 'wise' use rather than a 'beneficial' use (as Ciriacy-Wantrup puts it). This means that farmers can use groundwater not for paddy, not for sugarcane, but for other low water—high value crops, and not through flood or furrow or surface irrigation but through micro or drip or sprinkler irrigation.

Step 6: The government has virtually no scheme or programme towards irrigation extension. They only have agriculture extension by the agriculture/horticulture departments. Even agri-extension efforts are dwindling since 'Gramsevaks' have virtually vanished from villages of Karnataka and India. The government can make use of our agri-engineering graduates including diploma in agri-engineering graduates by offering them hands on training in groundwater, linking them with the Department of Mines and Geology which handles groundwater. The government may offer benefits from developmental programmes to those farmers who comply with disciplined groundwater use, such as subsidies from agriculture department/horticulture department and all other schemes. The disciplined groundwater use needs to be certified by the agri-engineering graduates/diploma holders who are registered as trained and needs to be cross checked by agri-department officials including the Department of Mines and Geology.

Step 7: India is the largest pumper of groundwater in the world, pumping twice the amount of the USA and 6 times that of Western Europe. India has the largest number of irrigation wells, namely, 25 million pumping around 75 ac-in. of water per well. At present, groundwater resource is meeting 80 % of the irrigation needs of India, while surface water meets the balance of 20 %. Even with the Department of Mines and Geology, Irrigation Department, Minor Irrigation Department, BWSSB, Urban Water Supply Board and all the laws such as Irrigation Act of 1965, Karnataka Land Reforms Rules of 1974 and the latest Groundwater Regulation and Control Act of 2011, the groundwater resource has been treated as an 'orphan' since there is no agency to practically conserve it (like the Forest Department which exclusively deals with forestry). The Department of Mines and Geology is more concerned with mines since minerals are more 'valuable' than 'groundwater'. The state should develop a body (though it has initiated the Groundwater Regulatory Authority, with teeth at ground level) to effectively address the groundwater resource use, similarly to how it addressed 'agriculture' during the 'Green Revolution', educating and creating awareness among farmers, measuring groundwater use by using water meters (not electrical meters) and helping farmers to understand the true value of groundwater so that they will use it efficiently, effectively and sustainably.

Reference

Chandrakanth MG (2009) Karnataka state water sector reform: current status, emerging issues and needed strategies, ITP Research series 3, IWMI-TATA water policy program, International Water Management Institute, ICRISAT, Patancheru, p 22. http://www.toenre.com/downloads/2009_11_Karnataka_water_sector_reform_IWMI_MGC.pdf

Chandrakanth MG, Nagaraja MG (2014) Payment for ecosystem services of water—case of Cauvery. Curr Sci 107(9):1375–1376

Chandrakanth MG, Romm J (1990) Groundwater depletion in India—institutional management regimes. Nat Res J 30:485–501. http://lawschool.unm.edu/nrj/volumes/30/3/03_chandrakanth_groundwater.pdf

Diwakara H, Chandrakanth MG (2007) Beating negative externality through groundwater recharge in India: a resource economic analysis. Environ Dev Econ 12:271–296. http://journals.cambridge.org/action/displayAbstract?fromPage=online&aid=942604&fileId=S1355770X06003500

Nagaraj N, Chandrakanth MG, Gurumurthy (1994) Borewell failure in drought prone areas of southern India—a case study. Indian J Agric Econ 49(1):101–106. http://www.toenre.com/downloads/1994_IJAE_NBD_application_paper_on_borewell_failire_NN_MGC_Gurumurthy.pdf

Chapter 6
Marginal Productivity of Water

6.1 Preamble

The concept of 'marginal' used in agricultural economics is the most interesting and crucial aspect used in production economics (PE). The concept of 'marginal' is actually observed and compared to 'average' or 'total'. The best example for a preliminary understanding of 'marginal' comes from the game of 'cricket', where each 'ball' bowled is observed to 'contribute' in terms of 'run/s'. The total product (TP) here is the total runs scored by each batsman and is the sum (or integral) of run/s scored from each marginal (numbered) ball (i.e. the first ball, second ball,...the first over, the second over,...) bowled by the bowler/s. The bowler bowls every numbered ball, that is, she/he bowls the first ball, the second ball of the first over...the first ball and second ball of the 10th over, etc., and correspondingly for each ball, the batsman can respond. Thus, marginal is the run/s scored for each ball which is observable. Upon adding (or integrating) the run/s scored for each (marginal) ball, the total runs scored is obtained. The average run scored per ball will then be obtained by dividing the total runs scored by the number of balls bowled. While the 'marginal' product highlights the true contribution of the batsman, the 'average product' (AP) masks the true contribution of the batsman especially how she/he contributed during crucial time/s of the game. For the selection of the batsman/batswoman, more so for one day match, the 'marginal product' (MP) is crucial compared to the AP. Given equal AP of two batsmen, the MP reflects also whether or how the batsman actually contributed to the success (failure) of the team in the game. For example, given two batsmen with the same average score (AP) per match, one who has contributed more to the success of the match by contributing during crucial times than the other batsman will be selected. In the ultimate analysis, if the AP of a batsman is higher than of anyone else, the contribution of another batsman at crucial junctures of the game which offered the 'turning point/s' towards success/failure (MP) will be relatively more important than the AP.

Taking another example of running race in the Olympics, the success of the runner is measured in terms of nanoseconds, rather than seconds. Therefore, when runners are completing their track, the finish is decided based on the particular nanosecond, which is a marginal concept. During the course of preparation for

M.G. Chandrakanth, *Water Resource Economics*, DOI 10.1007/978-81-322-2479-2_6

examination, if mother advises son to study for one hour between 4 am and 5 am, while father advises son to study any one hour, then mother is using the 'marginal' concept, while father is using the Average concept.

6.2 Average Product Influenced by Extreme Values

While the 'average' is influenced by extreme values, 'marginal' highlights the contribution of each input in the production process, the position of which is labelled keeping other inputs constant. Similarly, in the game of life, the teachers who taught us in the 'nursery' school, or the classmate/s or others responsible for our 'turning points' in life, are the MPs which are remembered by us forever in life. The greatest turning point is the inflection point on the production function, which describes the maximum of the contribution of a specific dose of input or the maximum MP. But geometrically, the inflection point is not as apparent as the maximum product on the production function (Fig. 6.1). What is to be noted is that there is an inflection point on the TP curve, which shows the maximum point on the MP curve. But the inflection point is not as apparent as the point of maximum TP. The inflection point is the point at which the slope of the TP is a maximum. This implies that the point where an input contributes its maximum gives the inflection point. Very few production functions, such as the transcendental model, accommodate the inflection point, and most other production functions, such as the Cobb Douglas model, the quadratic model, the square

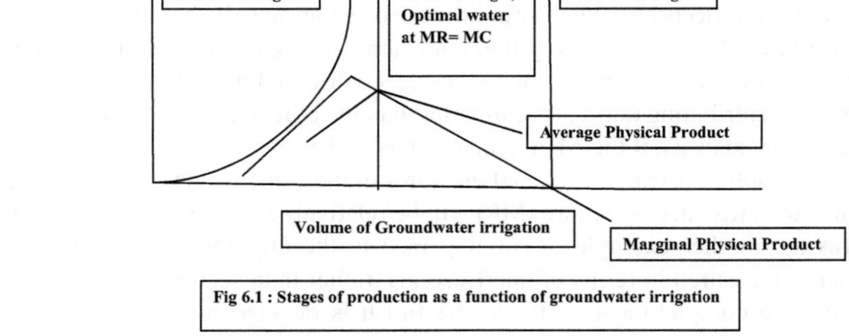

Fig. 6.1 Stages of production as a function of groundwater irrigation

root model, the Spillman–Mitscherlich model, do not accommodate the inflection point. The inflection point is of great significance for an input, since it provides the input level or dose at which the input's contribution is the greatest to the total output. For instance, this answers the question, which acre inch or hectare centimetre of water, the 10th or 15th or 23rd, contributes the most to the paddy output. The inflection point also gives another lead to the user that the farmer or the producer is still in the first stage of the production function and it is safe for him to increase the use of inputs further to realize higher output and profit. The stage of profit maximization is in the rational stage in the production function (Fig. 6.1). The maximum profit can be obtained in this region. But, if the input/s are fully subsidized and are virtually available free of cost to the farmer, it is profitable for the farmer to maximize the production of output as well as the profit. If the farmer has to pay market price for the variable input, his/her profit lies in rational stage in the decreasing returns stage, where the marginal productivity is falling, not when the MP is rising. If $Y = f(X)$, where Y = output, X = input, then profit Π = revenue − costs = $P_y * Y - (P_x * X + K)$, where P_y = price of output, P_x = price of input, K = fixed cost. Then for profit maximization $d\Pi/dx = 0$ implies $P_y * dy/dx - P_x(1) = 0$, which implies that for profit maximization $dy/dx = P_x/P_y$. This implies that the rule for profit maximization is the MP = $(dy/dx) = P_x/P_y$ = relative price ratio. Thus, if the farmer is able to match between the marginal productivity and the relative price ratio, then it is that specific level of X at which the MP = relative price ratio, and then we solve for X^*, which gives the optimum level of input, at which we get the optimum level of output Y^* as also the maximum profit.

6.3 Marginal Productivity

In simple terms, MP is the contribution of the specific speck (bit) of an input to the output, keeping other inputs constant. The crux of the MP is that MP is a precursor to the TP. Without MP, there is no TP, that is, without the contribution of an input, there is no output. Hence, the TP is originated from the MP and the AP is from the TP. As MP is the contribution of the specific speck (bit) of an input to the output, this contribution should be strictly 'observable'. Therefore, contribution of the 10th acre-inch of irrigation water to paddy, contribution of the 3rd acre of paddy land, contribution of the 18th man day of labour in the production of onion, contribution of the 518th rupee of variable capital to the output of milk are the MPs of irrigation water in paddy, land in paddy, labour in onion and capital in milk; these are theoretically observable. However, in agriculture, as these contributions involve the 'time' component, these marginals (or contributions) are not directly observable. The first kilogram of fertilizer applied during sowing may show its effect during intercultural operations. The 50th kg of fertilizer applied during interculture as the second dose may show up its effect during flowering and harvesting. The marginal productivity of water is measured in Chap. 9.

6.4 Division of Labour Is Impossible in Farming

In addition, the 'division of labour' is impossible in agriculture, and hence the farmer has to be the 'jack of all' trades in farming since farm operations are time consuming and operations have to be performed in a sequence over time. Thus, the farmer ought to know all the operations from land preparation, ploughing, sowing, interculture, fertilizer application, harvesting, threshing, on farm processing (if any), winnowing to sorting, grading, packing, storing, marketing, else she/he cannot sustain in agriculture. Thus, given these limitations imposed by land resource, the operation of the law of diminishing marginal returns (LDMR) and the 'time' involved in evoking response or contribution of the specific unit of input to the output, the MP is implicitly observed in agriculture.

Assume a farmer is cultivating paddy crop using surface irrigation from canal water. In the pool or canal of water, every acre-inch or hectare centimetre is uniform. Theoretically each acre-inch or hectare centimetre should contribute the same, to the output of paddy. However, as the operation of the LDMR sets in, with increase in water applied to a given area of (paddy) land, the MP of water goes on falling continuously from the point of inflection. The point of inflection shows that the dose of water which contributes the highest to the output of the paddy. After this inflection point, which sets in the LDMR, the marginal productivity of water goes on falling, and there is a point at which the MP is equal to the average productivity or AP of water. This point divides the water production function of the paddy into two parts, the first stage (or the irrational stage) from planting to the point where MP = AP and the second stage (or the rational stage) from the point MP = AP to the point where the MP of water is zero beyond which is the third stage or irrational stage. At this point, the paddy output reaches its maximum in which case the MP becomes zero and negative thereafter. Thus, the MP of each homogenous acre-inch or hectare centimetre of water varies at every stage of the production function due to the operation of the LDMR. Water applied initially contributes higher and higher to the paddy output, reaching a peak contribution called the point of inflection, and later on, as the LDMR operates, the MP falls. In clear terms, the MP measures the contribution of a specific speck (bit) of an input to the output. The term marginal is associated with marginal productivity or MP. Most text books define MP as the added output due to added input or the incremental output due to incremental input, not emphasizing 'keeping all other inputs constant'. Since the MP is measured in physical terms, it is also called marginal physical product.

6.5 Observed Reality and Estimated Utopian

While 'marginal' is an observed reality, 'average' is an estimated utopian. As 'total' is from 'marginal', and as 'average' is from 'total', 'average and total' form part of the 'history' or the 'past', while 'marginal' reflects the current or present situation.

Mathematically, the 'marginal' is integrated to obtain the 'total', and 'total' is divided over all units of input to obtain the 'average'.

6.6 Interpreting Marginal and Average Products

In the case of AP of labour, we express, for example, 0.15 quintal of wheat per man day of labour as the average productivity of labour. However, in the case of MP of labour, we express 0.15 quintal of wheat is contributed by the third man day of labour, and this is the marginal productivity of the third man day of labour. Thus, MP leads to decision making or normative economics concept, while AP is just an information or a positive economics concept.

6.7 Approach to Understanding the Marginal Product

There are conceptual and empirical differences in the MP as viewed in PE and farm management economics (FME). In PE,

Lt $\Delta Y/\Delta X = dY/dX$ (to be read as Limit of Delta Y/Delta X, as Delta X tends to zero = dY/dX) implies the MP given Y = TP and X = input. As $\Delta X \rightarrow 0$. As MP is the contribution of that 'speck (bit)' of input, which is due to $\Delta X \rightarrow 0$, this makes the size of incremental input 'X' so small to enable the measurement of its contribution, in PE. In 'farm management' however, the MP is the ratio $\Delta Y/\Delta X$, where the size of ΔX of incremental input can be of any magnitude. To repeat, in PE the size of ΔX is infinitesimally small, in order to obtain the contribution of a specific speck (bit) of input X, while in FME, the size of ΔX or incremental input can be any magnitude and is at least equal to one. In PE, theoretically the incremental input can be 0.1 or 0.001 or 0.0001 and so on. Conceptually, therefore, the definition of MP in farm management is more of an 'average nature', while that in PE is of a 'specific' nature.

Empirically, the definition of MP in farm management lies in the fact that the MP of any input is specific *to an individual farm*. Empirically, the definition of MP in PE is tied or bound by 'the functional form' of the 'production function' used, which is estimated using data from a sample of farmers. Thus, if a linear production function is estimated for the data of output and input from a sample of farmers, we get a constant MP, while if a CD production function is estimated for the data, the MP will be a variant of the elasticity of production with respect to both output and input. Similarly, the MP in PE varies not just with the empirical data but also with the 'model specification' unlike MP in FME. And therefore, the MP in PE offers a 'policy measure' since it reflects the contribution of a specific speck (bit) of an input considering the data from sample farmers, unlike the MP in FME. For an extension specialist, the FME definition of MP is crucial, while for a policymaker, the PE definition of MP is relevant.

6.8 The Case of Multiple Inputs

In the production process with multiple inputs, conceptually and empirically the definition of MP will have a sea of change. In the case of FME, the MP will just be $\Delta Y/\Delta X_1$, $\Delta Y/\Delta X_2$,…$\Delta Y/\Delta X_i$. This definition conceptually overestimates the MP since the numerator is the same for all the inputs and this incremental output ΔY is just arithmetically distributed across all units of added inputs ΔX_i, without any ceteris paribus conditions, as in the PE definition. Thus, the FME definition of MP is not only an 'average' measure but also an 'overestimate' of the input contribution to the output. In PE, in the case of multiple inputs, the MP will then be the partial derivative of the output with respect to each input which means, the partial derivate of output is taken with respect to the specific input, keeping all other inputs constant. The definition of MP will then be the contribution of the specific speck (bit) of an input to the output, keeping all other input uses constant at their mean level. Whether the mean level is arithmetic or geometric or any other mean depends upon the model specification of production function Y.

6.9 Understanding Ceteris Paribus or Keeping Other Things Constant

The Latin phrase ceteris paribus is the key to the understanding and mathematical interpretation of partial marginal productivity in the context of production function and arises due to the use of multiple inputs. Accordingly, the marginal productivity of an input is influenced by the marginal productivity of other input/s. For instance, the MP of irrigation water is influenced by marginal productivity of the fertilizer and vice versa.

In a physical limited sense, it is in fact difficult to *keep other things constant* as meant by ceteris paribus. This applies to both the LDMR and the law of diminishing marginal utility (LDMU). In LDMU, the marginal utility of a good falls with the increase in its consumption, keeping income, taste and preference, prices of related goods constant or ceteris paribus. In reality, none of these variables can be kept constant, for example, the taste and preference. If a student is eating a nice food in the cafeteria, when another student looks at him/her, it motivates him/her to buy the specific food item as well, and thereby increases its consumption. Hence, keeping other factors constant as meant by ceteris paribus needs to be understood as *in the presence of other variables which are treated as constants for the purpose of partial derivative* but *not as in the absence of other variables*. Similarly, incomes can never be constant nor prices of related goods. In the law of demand,[1] the quan-

[1] The same is true in the case of 'law of demand' or 'law of supply' where ceteris paribus implies keeping the 'demand shifters' or the 'supply shifters' constant respectively. In reality, neither the demand shifters nor supply shifters can be kept constant. Thus the consumer behaviour is observed in the presence of demand shifters such as family income, family size, prices of substitutes/complements, advertisement expenditure, subsidies, taxes. The producer behaviour is observed in the presence of supply shifters such as technology, weather, prices of substitutes/complements, subsidies and taxes.

tity demanded of a commodity varies inversely with price; ceteris paribus can only mean that the quantity demand of a commodity varies inversely with price, in the presence of other variables such as income of the consumer, prices of related goods, tastes and preferences, which are 'treated' as constants, for the purpose of obtaining the 'partial' response of the quantity demanded due to the change in 'price' of the commodity in question using the partial derivative of the demand function.

In LDMR, how can we keep 'other input uses constant at mean level'? In reality, nothing can be kept constant in the field level. The farmer will apply inputs singularly and simultaneously, without holding anything constant, except for the size of his own land and a few other fixed inputs. *Thus, holding other inputs constant in reality implies 'in the presence of other inputs' at their mean level*. Thus, the mathematical definition of partial marginal productivity does not give a 'feel' for the understanding of 'partial' marginal productivity since nothing can be held constant in reality (Table 6.1). As an illustration, in a production function of sugarcane with two inputs, chemical fertilizers and irrigation water, the MP of chemical fertilizer and the MP of irrigation water can be computed in PE by taking the first derivative

Table 6.1 Marginal product (MP) in farm management economics (FME) and production economics (PE)

MP	FME	PE
Definition	Added output due to added input	Contribution of a specific speck (bit) of an input to the output
Basis	Farm level data for specific farm	Estimation of production function using field data from farmers is essential for estimation of the MP
Formula	$\Delta Y/\Delta X$	$\Delta Y/\Delta X = dy/dx$ for single input as $\Delta X \rightarrow 0$ $\Delta Y/\Delta X = \delta y/\delta x$ for multiple inputs as $\Delta X \rightarrow 0$
Expression	Contribution of a specific input of input to the output, e.g. contribution of the second man day of labour in wheat cultivation is 0.15 quintal of wheat	Contribution of a specific input of input to the output, e.g. contribution of the second man day of labour in wheat cultivation is 0.15 quintal of wheat, keeping all other inputs constant (ceteris paribus)
Status	Marginal productivity is a discrete measure	Marginal productivity is a continuous measure
Limitation	Specific to a farm in a micro-sense and hence bound by a farm situation	Specific to a group of farmers in a macro-sense and hence bound by the model of the production function
Conceptual	Average over a wide input range	Specific for a speck (bit) of an input
Relevance	Relevant to individual farm	Over a range of farms, hence useful in policy-making
Multiple inputs	MP is an overestimate since it is an arithmetic division of incremental output by incremental input, without relevance to other input/s	MP is a partial marginal productivity, where the contribution of a specific speck (bit) of an input is measured in the presence of other input/s at mean level

of a production function. Accordingly, the MP of the 10th kg of a chemical fertilizer keeping irrigation water constant can be 0.25 t of sugarcane. But the interpretation should be that the MP of the 10th kg of chemical fertilizer = 0.25 t of sugarcane in the presence of irrigation water since irrigation water or any other variable can never be kept constant. Thus, in the case of a quadratic model with interaction term, the MP of a chemical fertilizer includes the MP of irrigation water and vice versa. On the other hand, if we interpret ceteris paribus as keeping other variables constant, the concept of MP can neither be understood nor appreciated since nothing can be kept constant in reality, as the MP of one variable input is in fact influenced by the presence of other input/s. In FME, these nuances cannot be captured.

Chapter 7
Costing Water for Irrigation

7.1 Properties of Water as a Natural Resource

1. Water resources are unique and accordingly have no substitutes. The economic implication is that their elasticity of substitution is zero, which implies that water resources have to be best used as complements and in addition have absolutely no substitute or alternative. Groundwater in areas where river is absent is an unique natural resource with virtually no substitute.
2. Uncertainty of water resources in stock/flow as well as the human factors such as efforts to recharge shapes the cost of water resource. Growing uncertainty in the availability of groundwater over years reflected in the increasing probability of initial as well as premature well failure exacerbates the cost of groundwater.
3. Irreversibility of water/groundwater resources implies that once they are mined/misused/overused, it is difficult to regain the original position or to be resilient. Hence, transaction costs of resilience are colossal and result in increasing investment on wells due to increase in premature/initially failed wells. Similarly, if the sources of natural streams, springs of water, are affected, they cannot be rejuvenated forever or the cost of rejuvenation is colossal.
4. Indispensability of water resources implies that equity and sustainability in use of water/groundwater as a natural resource is crucial. Hence, proper accounting and valuation of groundwater is crucial to impress upon the users and uses regarding the cost/value of groundwater, even though the resource lacks well-defined property rights.
5. Invisibility of water resources imply that these resources are sometimes visible only upon extraction and sometimes continue to be invisible as the streams have merged themselves into other streams (e.g. the Saraswathi river is often called Gupta Gamini in Triveni Sangama or Prayag at Allahabad, the confluence of Ganga, Yamuna and Saraswathi), hence difficult to make predictions regarding their stock and flow. Groundwater from borewells/tubewells is a classic example of invisibility, which has pervasive influence on its cost/value.
6. Water resources are remote and location specific—as in remote locations—as, for example, fossil groundwater, and the remote location imposes costs in

M.G. Chandrakanth, *Water Resource Economics*, DOI 10.1007/978-81-322-2479-2_7

extraction and transportation (incorporated in drilling, casing, conveying costs). There are instances where farmers have spent more on transportation than on drilling, as groundwater was not available on their farm or in the proximity.

7. Water resources are intricate—as, for instance, the properties of aquifers are subject to the nature of rock formation (hydrogeology) and the degree of weathering, which result in confined/unconfined aquifers. In hard rock areas, groundwater availability is intricate as also interdependent and synergistic—rather than independent and divisible. Hence, costing cannot be independent of factors contributing to their formation—especially groundwater formation depends on the nature of the aquifer—hard rock vurses alluvial, which have different properties. Hence, the cost of groundwater will obviously be higher in hard rock aquifers than the alluvial aquifers.
8. Water resources, especially groundwater from wells, are fraught with externalities, where the action of a farmer through drilling a well has unintended side effects on the other farmers, and this effect influences the production/consumption function and results in inefficiency and welfare loss, as there is no price mechanism or contract governing the externality. In the case of groundwater, due to violation of isolation (interwell) distance, farmer drilling borewell(s) have an unintended side effect on neighbouring farmers' well/s, resulting in the increase in cost of extraction due to the increase in the number of well/s drilled, inefficiency and welfare loss. Further, as there is no mechanism to compensate for the loss inflicted and no institution governs the behaviour of farmers, the externality persists. This is a classic case of reciprocal externality, as it is difficult to identify which well is responsible for over extraction and/or failure of neighbouring well/s similar to the traffic pollution externality. Unless there is a regulation/institution on extraction, use, recharge and reuse, the externalities persist, resulting in inefficiency as well as welfare loss.
9. Time of availability of the natural resource: In the case of groundwater, the shrinking life/age of wells, increasing drilling depths and costs, reduced groundwater output, increasing pumping costs, increased pump capacities, associated costs of micro-irrigation are posing challenges in valuation. For example, groundwater in open wells/tubewells used to serve for 30–40 years are now serving for 2–3 years or even less. Open wells or dugwells have become defunct in many HRAs. Thus, depreciation on fixed investment is no longer relevant since investment on digging or drilling and casing is becoming a variable cost. Thus, amortized investment on both functioning and nonfunctioning wells (due to externalities) enter as the marginal cost influencing decision-making rather than fixed cost since *depreciation is a mere replacement cost not used in the decision-making process.*

Paradoxically, even with the innumerable number of organizations on water—such as the Central Water Commission, Ministry of Water Resources, Central Groundwater Board, National Water Development Authority, state water resource departments, state departments of mines and geology, urban and rural water supply development boards—efforts towards volumetric measurement of water applied are still crude and approximate. In fact virtually there is no measurement of surface water or groundwater resource and even though Agriculture students undergo the course on water management, they seldom apply their knowledge in the field. Thus, water cost is not properly accounted in any of the costing procedures including the Commission on Agricultural Costs and Prices (CACP), which have no adequate information on water use in the Record Type (RT) forms.[1] Therefore, there are no compelling reasons to accept that the costs of cultivation and the minimum support price (MSP) are properly estimated, and they are more towards underestimation. The CACP computes depreciation of investment on well and accessories over an unspecified number of years (subjective and left to the discretion of the field assistant who obtains field data from farmers using the cost accounting method). The depreciation is an outmoded concept in costing groundwater irrigation, due to the presence of reciprocal negative externality, which is increasing cost of groundwater.

7.2 Hitherto, Why Cost of Water/Groundwater Was Not Accounted in Irrigation

7.2.1 Surface Water

In the era of Green Revolution and earlier, water in general and groundwater in particular did not relatively exhibit signs of physical and economic scarcity in the hard rock areas as well as in alluvial basins. Perhaps, administrators as well as researchers did not pay adequate attention to economic costing of irrigation water as surface water and groundwater supplies did not pose any scarcity threats. The cost of cultivation for irrigated crops did not accordingly include the cost of water used in irrigation. At the most, the cost of irrigation was equivalent to the cost of labour involved in irrigating crop/s. In fact many studies dealing with economics of irrigation at the most considered frequency of irrigation to represent volume of irrigation, which is fundamentally wrong. But, the data on frequency of irrigation does not reflect the volume of water applied to a crop since the frequency of irrigation is invariant with the area irrigated. Thus, research articles appearing in standard economic journals such as the Indian Journal of Agricultural Economics, Economic Political Weekly during the period of Green Revolution till date did not properly portray the cost of irrigation water as a separate entity. In articles dealing with surface water, 'water rate' announced by the respective state governments have been used to reflect the cost of surface water. However, the water rate is a linear measure of water and not a volumetric measure. For instance, a farmer who pays a water rate of ₹ 100/acre of paddy in the tail reach of a command uses applies much

[1] The RT 440 of CACP has the information pertaining to the type of well, number of wells, HP of pump, command area irrigated, percentage owned, year of drilling, age at present, remaining life, amount invested, value at present and salvage value. However, there is no information on the expected age or life of wells; it is assumed to be 10 or 20 years and is left to the imagination of a field assistant who collects the data. RT 441 deals with the change in a well and indicates when the well was destroyed (or failed) and when a new well was constructed. There is no information on the volume of groundwater yield of well/s extracted by farmers.

lower water than the farmer located in the head reach, and pays the same water rate. In addition, the fixed water rates are not based upon economic costing approaches as they are more of an institutional nature.

One of the methods of economic costing of surface water is illustrated here. Consider the investments made on the construction of an irrigation dam. The yearwise investment on the construction of the reservoir or dam is obtained. This investment is compounded from the year of investment to the present year at a social interest rate of 2 %.

The choice of a low discount rate of 2 % is debatable and is discussed in detail in Diwakara and Chandrakanth (2007). Considering the vintage of borewells drilled at different years where the well depth ranged from 80 to 150 ft in 1985 doubled in 2001, while the yield dwindled. Simultaneously, pump capacity increased from 5 to 7.5 HP (horsepower) along with the number of stages of pump due to increased drilling depth. For instance, investment per well increased from ₹ 53,605 to 74,190 in two decades, which approximates to 2 % compound growth rate. Thus, the rate of increase in nominal investment in irrigation wells can be taken around 2 % compound growth rate. The authors indicate that the real growth of investment in irrigation wells is negative, which needs further validation. Thus, investment on irrigation wells growing at 2 % offered a close approximation to the growth of investment in irrigation wells. The choice of 2 % in the rate of investment per well is reasonable, but, if externalities are included, this may increase.

The yearwise compounded cost of dam construction obtained from the initial year of investment till the year under consideration (say 2014) is summed up to obtain the total compounded cost. The total compounded cost is amortized over 100 years, assumed as the life of the irrigation reservoir project at 2 % to obtain the amortized investment per year representing the fixed cost. The yearly operation and maintenance costs are added to the fixed cost to obtain the total cost of water. This total cost of water per year is divided by the yearly water outflow from the reservoir/dam for irrigation to obtain the cost of water on volumetric basis per acre-inch or per hectare centimetre. Nagaraj et al. (2003) used this methodology and found that that the cost of surface water is ₹ 12/acre-inch on volumetric basis or ₹ 1007/acre on area basis. Considering for 2014, this cost works to ₹ 1252/acre on area basis and ₹ 15/acre-inch on volumetric basis incorporating the inflation rate. For paddy, with the governmental water rate being ₹ 100/acre, the subsidy amounts to ₹ 1152/acre of paddy or subsidy of 92 %. Similarly, for sugarcane the subsidy amounts to ₹ 852/acre or subsidy of 68 %. Thus, the water rate fixed by the Government of Karnataka, viz. ₹ 100/acre for paddy, ₹ 400/acre for sugarcane, ₹ 35/acre for semidry crops (like ragi, jowar, groundnut) payable to the revenue department is highly subsidized, and farmers do not recognize their responsibility to pay these water rates.

7.2.2 Groundwater

In the case of farmers using groundwater, earlier studies treated the investment on wells/borewells, irrigation pump set and conveyance structures (CSs) as fixed cost.

However, in hard rock areas, the life and age of irrigation wells are fast reducing due to overexploitation of the resource in the post-Green Revolution period. The post-Green Revolution can be called as groundwater overextraction revolution. As the farmers do not stick to maintaining the isolation distance between borewells, also due to small-sized holdings, further subject to subdivision and fragmentation of holdings, borewells are mushrooming. The dug wells which were yielding sumptuous water during the Green Revolution virtually began to dry up, due to overexploitation of groundwater from nearby dug wells and shallow borewells. Shallow borewells began to dry up due to overexploitation from deeper tubewells. Thus, due to reciprocal externality (explained in the chapter on externality), the probability of well failure is increasing, and simultaneously the life/age of the borewell is falling along with the water yield of the borewell. Under such circumstances, the cost of groundwater will increase reflecting both physical and economic scarcity.

The life/age of an irrigation well is specific to each farm in hard rock areas. Here, factors inter alia, such as aquifer characters, degree of cumulative interference, efforts to recharge irrigation wells, electricity supply, markets, road connectivity, availability of labour and institutional factors such as sharing water shape the economy of groundwater irrigation. During the Green Revolution period, the irrigation wells/borewells used to function for periods exceeding 20–30 years due to extraction of groundwater almost matching with recharge, and the water extraction mechanisms were manual lifts, such as Yetha, Picota and Persian wheel (bucket machine). In the 1970s, during the Green Revolution period, due to use of sustainable extraction lifts and sustainable extraction of groundwater-supported cropping systems, the life and age of the wells sustained 10–20 years or more. Hence, the cost of cultivation studies conducted during the Green Revolution and post-Green Revolution period considered the cost of groundwater as fixed cost, which did not influence the decision-making ability of farmers. Studies conducted on economics of irrigation including CACP have little consideration for the cost of water/groundwater.

7.3 Concept of Initial Failure, Subsistence Life, Premature Failure and Economic Life of Wells

Initial failure of borewell refers to a borewell which did not yield any groundwater at the time of drilling and thereafter. Subsistence life of borewell refers to the number of years a borewell yielded groundwater for the payback period (PBP). The PBP is obtained by dividing the sum of the total investment on drilling, casing, IP set, CS, storage structure, drip/sprinkler structure, recharge structure and electrification charges of the borewell by the annual gross returns from irrigation borewell. The hypothesis is that an irrigation borewell is considered to have served its purpose if it has at least paid back the total investment made for the purpose. This implies that the PBP indicates the period in which a borewell recovered the investment made.

Premature failure refers to the borewell which served below the subsistence life or the PBP. Economic life/age of the borewell refers to the number of years a borewell yielded groundwater beyond the PBP.

The crop pattern, irrigation intensity, technology of irrigation adopted, availability of electricity and economic factors determine the demand side of groundwater. With the myopic behaviour of farmers towards extraction (draft) and recharge, given the asymmetric information and unsustainable extraction processes, the definition of *economic life of the borewell* is crucial.

7.4 Costing Groundwater in Irrigation

In the period after 1980, and more recently after 1990, due to the increasing rate of initial and premature failure of borewells, investment on drilling and casing of irrigation wells which was hitherto considered as fixed cost can no longer be considered as fixed cost but instead needs to be considered as *variable cost*, as the increasing marginal cost of groundwater enters decision-making. Thus, annual cost of irrigation comprises the variable cost component and the fixed cost component. The variable cost of groundwater is obtained by amortizing the investment on drilling and casing of borewells over the subsistence life of borewell/s or economic life of borewell/s (whichever is relevant for the specific borewell) plus the operation and maintenance costs of the borewell. The amortized investment is divided by the volume of groundwater extracted in the recent year of collecting field data to obtain the variable cost of groundwater per acre-inch.

The fixed cost of groundwater is obtained by amortizing the investment on irrigation pump sets, pump house, electrification charges, groundwater storage structure (constructed if any), groundwater delivery pipe investment, drip irrigation and accessory investment for a period of 10 years since these structures can be assumed to serve for at least a decade. The amortized fixed investment is divided by the volume of groundwater extracted in the recent year to obtain the fixed cost of groundwater per hectare centimetre or acre-inch.

The fixed cost of groundwater recharge structure is obtained by amortizing the investment on groundwater recharge over the subsistence or economic life of borewell, whichever is relevant for the borewell. The labour cost of irrigation is considered along with labour costs of other cultural operations. The annual cost of irrigation pertains to each irrigation borewell on the farm and is added across all borewells on the farm. This total cost of irrigation is then computed for each crop according to the volume of groundwater used in each crop. Cost of irrigation per acre-inch or ha cm = [total annual cost of irrigation]/[volume of water used for the crop] in acre-inches or ha cm of groundwater used.

7.4.1 Amortized Cost of Borewell

As mentioned earlier, inorder to obtain the annual share of groundwater irrigation cost, the borewell investment needs to be amortized with the assumption that investment on drilling and casing is no longer a fixed cost, given the increasing probability of well failure forcing farmers to continue to make investments to irrigate crops through new borewells/drillings. In many areas of hard rock areas, the probability of borewell failure is around 0.7. In cases where farmers are not showing tendencies of overexploitation, the probability of well failure drops to 0.3. Thus the magnitude of probability of well failure in itself reflects that for most farmers, investment on borewell exploration equal to the cost of drilling and casing renders a variable cost rather than fixed cost. This investment is amortized (or apportioned with interest) over the average life of the well. The amortized cost varies with the amount of capital investment, age of the borewell, discount rate and the year of the construction of the borewell. Amortized cost of the borewell was worked out as under:[2]

Amortized cost of irrigation = (amortized cost of borewell + amortized cost of pump set + amortized cost of conveyance + amortized cost of overground structure + annual repairs and maintenance costs of pump set and accessories) given by:

$$\text{Amortized cost of BW} = (\text{compounded cost of BW}) \times \frac{(1+i)^{AL} \times i}{(1+i)^{AL} - 1]},$$

where

AL= average age or life of borewell i = discount rate considered = 2 %.

$$\text{Compounded cost of BW} = (\text{historical investmenton BW}) \times (1+i)^{(2014-\text{year of drilling})}$$

if 2014 is considered as the latest reference year.

$$\text{Amortized cost of P and A} = (\text{compounded cost of P and A}) \times \frac{(1+i)^{10} \times i}{(1+i)^{10} - 1]}$$

The working life of pump sets and accessories (P and A) is considered to be 10 years since farmers consider 10 years as their economic life.

i = discount rate considered at 2 %

$$\text{Compounded cost of pump set and accessories} = (\text{historical cost of P and A}) \times (1+i)^{(\text{say } 2014-\text{year of installation of P and A})}$$

$$\text{Amortized cost of conveyance structure (CS)} = (\text{compounded cost of CS}) \times \frac{(1+i)^{10} \times i}{(1+i)^{10} - 1]}$$

The working life of conveyance structures (CS) is also considered to be 10 years. The usual mode of conveyance of groundwater is through PVC pipe.

i = discount rate considered at 2 %

[2] Patil 2014.

$$\text{Compounded cost of CS} = (\text{historical cost of CS}) \times (1+i)^{(2014-\text{ year of installation of CS})}$$

$$\text{Amortized cost of micro irrigation structure} = (\text{compounded cost of MIS}) \times \frac{(1+i)^{10} \times i}{(1+i)^{10} - 1]}$$

The working life of micro-irrigation (drip) structure (MIS) is considered to be 10 years since farmers usually replace them after 10 years.

i = discount rate considered at 2 %

$$\text{Compounded cost of MIS} = (\text{historical cost of MIS}) \times (1+i)^{(2014-\text{year of installation of MIS})}$$

As a coping mechanism to endure with the persistent problems imposed by variations in supply of voltage in electricity to run irrigation pumps and supply of electricity during off-peak load hours and low yields of borewell, farmers have built overground storage structures. The amortized cost of overground storage structure is estimated as under:

$$\text{Amortized cost of overground storage structure} = (\text{compounded cost of OSS}) \times \frac{(1+i)^{10} \times i}{(1+i)^{10} - 1]}$$

i = discount rate considered at 2 %

$$\text{Compounded cost of OSS} = (\text{historical cost of OSS}) \times (1+i)^{(2014-\text{year of construction of OSS})}$$

$$\text{Amortized cost of borewell recharge structure} = (\text{compounded cost of BRS}) \times \frac{(1+i)^{AL} \times i}{(1+i)^{AL} - 1]}$$

AL = average life/age of a borewell

i = discount rate considered at 2 %

$$\text{Compounded cost of borewell recharge structure BRS} = (\text{historical cost of BRS}) \times (1+i)^{(2014-\text{year of construction of BRS})}$$

A modest discount rate of 2 % is considered for amortizing the cost of an irrigation borewell to represent the compound rate of interest in the costing borewell components such as construction cost, drilling cost, pump set, accessories, micro-irrigation structure and recharge structure. It is crucial to note that accessories in irrigation depends upon the farm endowments. If farmer has drip irrigation, s/he may not have groundwater storage structure for instance.

7.4.2 *Yield of Irrigation Borewell*

The field measurements of the groundwater yield of borewells were made by recording the number of seconds taken to fill a bucket or overground structure of groundwater of a known volume. Initially the borewell was put on for 10 min so that the initial pump yield bias is avoided. This was linearly extrapolated to obtain the groundwater yield in gallons per hour (GPH). In the case of lack of power supply at the time of data collection, the yield of the borewell was recorded, as perceived by farmers, as 1 in. = 1000 GPH, 2 in. = 2000 GPH and 3 in. = 3000 GPH and so on.

7.4.3 *Groundwater Use in Conventional Irrigation System*

The acre-inches (or hectare centimetre) of groundwater used for each crop in each season (summer, kharif, rabi) in the conventional system of irrigation is calculated as = [(area irrigated in each crop) × (frequency or number of irrigations per month) × (number of months of crop) × (number of hours for one irrigation for the cropped area in question) × (average yield of borewell in gallons per hour)]/22,611 which gives the groundwater use for each crop in acre-inches or ha cms.

7.4.4 *Groundwater Use Measurement in Drip Irrigation System*

The groundwater used for irrigation in each crop (acre-inches) in drip irrigation = {number of drips or emitters for the cropped area × groundwater discharged per emitter per hour (liters per hour) × no. of hours to drip irrigate the cropped area for one irrigation × frequency of irrigations per month (in number) × duration of crop irrigated in months/4.54/22,611}.

The groundwater used for irrigation in each crop (acre-inches) in sprinkler irrigation = {number of sprinklers for the cropped area × no. of hours to irrigate the cropped area for one irrigation × groundwater discharged per sprinkler (in liters per hour) × frequency of irrigation per month (in number) × duration of crop irrigated in months/4.54/22,611}.

One acre-inch is equivalent to 22,611 gallons or 3630 ft^3 and 1 ft^3 is equivalent to 28.32 L. Total groundwater use per farm is total acre-inches of groundwater used in all seasons across all crops including perennial crops.

7.5 Annual Cost of Irrigation

Farmers using irrigation pump sets (below 10 HP capacities) for groundwater are not charged for electrical power. The government (of Karnataka) however imposed a flat charge of ₹ 300/HP per year up to 10 HP pump set since April 1997. The power transmission corporations/state governments have been soft towards seeking electricity dues from farmers for the reasons of political economy. Hence, there are no explicit payments towards electricity for pumping groundwater, other than annual operation and maintenance charges of the irrigation pump set and borewell up to 10 HP.

However, the implicit cost of irrigation is relevant for farmers in hard rock areas due to the high probability of initial and premature borewell failure, which forces farmers to invest in additional borewell(s) to at least remain on the original production possibility curve. The investment on failed borewells is increasing due to violation of isolation distance between irrigation borewells, overextraction or mining of groundwater, lack of efforts to recharge groundwater and reciprocal negative externality. The resulting transaction costs are due to forced investment on drilling and casing of additional borewells since borewells drilled failed initially or prematurely to yield groundwater. The cost of groundwater varied from ₹ 200 to 2000/acre-inch, depending upon the degree of externality, the variable and fixed cost incurred on irrigation wells and the probability of initial and premature failures.

7.5.1 Externality Defined

The externality per well = (amortized investment on drilling and casing of borewells over the subsistence life of well/s or economic life of well/s whichever is relevant)/(number of wells which served PBP + serving economic life) − (amortized investment on drilling and casing of borewells over the subsistence life of well/s or economic life of well/s whichever is relevant)/all the wells on the farm.

7.5.2 Illustration of Cost of Groundwater Irrigation

Using the relevant formulae for variable and fixed costs of groundwater, the cost of groundwater is illustrated in Tables 7.1 and 7.2. The variable cost of groundwater irrigation ranges from 11 to 35 % of the total cost of cultivation of crops for seasonable vegetable crops in Karnataka, while the fixed cost of groundwater irrigation ranges from 2 to 14 % of the total cost of cultivation of crops. By excluding the cost of groundwater irrigation, the overestimation of net returns ranges from 18 to 94 % across vegetable crops. Thus, farmers need to be educated regarding the need to account for groundwater cost which they are already incurring. These costs do not include the pumping cost of electricity, and if the energy cost is included, the cost of cultivation will further rise. A preliminary estimate is around 100 kW h to lift 1 ac-in. of groundwater, which if valued at ₹ 3.5/kW h works to ₹ 350/acre-inch. For carrot, the energy cost would be ₹ 2657 per acre, while that for potato would be ₹ 4172 per acre. The groundwater cost is illustrated for perennial crops in Table 7.2. The percentage of overestimation of net returns in perennial crops is modest ranging from 5 to 19 %.

Table 7.1 Economics of vegetable crops including groundwater cost of irrigation on farms using drip irrigation in Karnataka

Crop	Groundwater used in acre-inches	Variable cost of groundwater (% of total cost)	Fixed cost of groundwater (% of total cost)	Total cost of cultivation per acre	Yield of crop per acre	Gross returns per acre	Net returns per acre including groundwater cost	Net returns per acre excluding groundwater cost	% of overestimated net return due to exclusion of groundwater cost
		₹	₹	₹	Qtl	₹	₹	₹	%
Coriander	4.70	11,765 (23)	7328 (14)	52,047	150	75,000	22,953	42,046	83
Potato	11.92	25,778 (24)	762 (0.72)	106,168	227	211,012	104,844	131,384	25
Cabbage	10.05	24,045 (18)	2304 (1.7)	135,310	230	230,476	95,166	121,515	28
Knol kohl	12.08	22,324 (35)	3776 (6)	63,002	155	90,666	27,664	53,764	94
Tomato	12.16	20,840 (14)	2107 (1.44)	146,044	10.50	238,689	92,645	115,592	25
Beans	10.31	25,944 (23)	4251 (4)	112,176	7.40	182,500	70,324	100,519	43
Capsicum	8.18	17,583 (13)	6067 (4.5)	134,400	4.50	180,000	45,600	69,250	52
Red onion	9.32	19,034 (27)	5625 (8)	71,019	96	136,693	65,674	90,333	38
Cauliflower	8.54	7321 (11)	2308 (3.5)	64,990	14,545	118,182	53,192	62,821	18
Carrot	7.59	17,349 (26)	2120 (3.1)	68,007	109	108,571	40,564	60,033	48

Table 7.2 Economics of perennial crops including groundwater cost of irrigation in Karnataka

Crop	Groundwater used in acre-inches	Variable cost of groundwater	Fixed cost of groundwater	Total cost of cultivation	Crop yield	Gross returns	Net returns including water cost	Net returns excluding water cost	% of overestimated net return due to exclusion of groundwater cost
		₹	₹	₹	Qtl.	₹	₹	₹	
Pomegranate	11.46	9087	154	131,583	2553	217,982	86,399	95,640	11
Papaya	15.25	9359	189	94,598	127.14	145,476	50,878	60,426	19
Banana	36.24	4729	213	61,915	44.42	157,121	95,206	100,148	5
Arecanut	12.16	4910	285	58,003	7.98	116,726	58,723	63,918	9
Coconut	8.41	2490	355	28,865	4880	57,600	28,735	31,580	10
Mango	12.36	2775	221	57,462	28.90	105,957	48,495	51,491	6
Sapota	12.03	2281	217	58,018	101.78	96,428	38,410	40,908	7

7.6 Conclusions

The cost of groundwater irrigation which forms around 11–35 % of the cost of cultivation of seasonal crops is an important economic cost. This cost is being ignored by all farm management studies as well as the Directorate of Economics and Statistics, which supplies information on cost of cultivation of crops to the CACP, the top policy-making body regarding fixation of the MSP. Accordingly, the present procedure for recommending the MSP excludes the cost of groundwater irrigation, and hence the farmers are not getting a fair/remunerative price, as the costs are underestimated to the tune of the cost of groundwater and overestimated to the tune of 18–94 % (in the case of vegetable crops). It is therefore crucial to economically account for the cost of groundwater/surface water in the cost of cultivation of crops.

References

Diwakara H, Chandrakanth MG (2007) Beating negative externality through groundwater recharge in India: a resource economic analysis. Environ Dev Econo 12:271–296 (Cambridge University Press)

Nagaraj N, Shankar K, Chandrakanth MG (2003) Pricing of irrigation water in cauvery basin, case of Kabini command. Econo Polit Wkly 38:4518–4520

Patil KKR (2014) Economics of coping mechanisms in groundwater irrigation: Role of markets, institutions and technologies. Unpublished PhD thesis, submitted to the Department of Agricultural Economics, University of Agricultural Sciences, Bangalore

Chapter 8
Locating Interference and Valuing Water

Interference among irrigation wells is a rule rather than an exception in hard-rock areas (HRAs). The main focus is to value the groundwater for irrigation in areas fraught with cumulative interference among irrigation wells. Explicitly this is to measure in economic terms, the effect of well interference (local overdraft) when pumping by a farmer affects the pumping conditions of his/her neighbour. This effect of interference may be localized in nature and may not be associated with a general decline in the water table. Initial or premature failure of irrigation well/s due to cumulative interference among wells occurs when farmers' extractions are affected by the pumping cones of other farmers. In this study, the well failure due to cumulative interference is defined as follows:

(i) The well dries up because of new well(s) coming in (but not because of decline in rainfall).
(ii) The well loses a large degree of its yield because of new well(s) coming in (but not because of decline in rainfall).
(iii) The well is deepened because of new well(s) coming in.

The well interference problem poses serious threats to the efficiency, equity and sustainability of well irrigation. A large farmer who can afford to dig, drill or deepen one or many irrigation wells can seriously hamper the irrigation prospects of a neighbouring small farmer who is irrigating with one or a few wells. In the long run, the small farmer may be forced to shift his/her operations to dry-land agriculture, which is a clear equity issue. The valuation of the effects of interference provides valuable information regarding the predicament faced by farmers due to well interference, which is largely induced by human actions and hence retractable if there is a social will. The challenge, however, is to locate the patches where interference is predominant.

M.G. Chandrakanth, *Water Resource Economics*, DOI 10.1007/978-81-322-2479-2_8

8.1 Methodology to Identify/Locate Interference Among Irrigation Wells

In Karnataka, even with modest support to minor irrigation by institutional finance after the mid-1980s, the number of irrigation wells is increasing despite the increasing probability of initial and premature well failures. Farmers have found that groundwater irrigation is the major mode of investment for farm development in different parts of the state. These farmers have also diversified their cropping to cover different risks. Such investments in groundwater irrigation wells are recurring in agricultural areas where investment on irrigation wells is already in vogue despite interference among wells. In part, the rationale for this investment behaviour is offered by the permanent income hypothesis of Friedman and relative income hypothesis of Duesenberry.

But, at the micro (farm) level, farmer A may/may not hesitate to locate the well/s which is/are responsible for interfering with his/her irrigation well. In case farmer A perceives that a specific well belonging to farmer B is causing interference, farmer B may/may not accept the same. Even if farmer A wishes to prove the effect of interference by using chemical or colour tracers, farmer B may/may not accept. Even if farmer B accepts that his well is responsible for interference, no action is called for, due to ambiguity in property rights. Groundwater being an invisible, irreversible and fugitive slowly renewable resource exacerbates the predicament.

Even at the macro (village/taluk) level, no relevant statistics, either directly or indirectly, reflecting the interference among wells are reported by the Directorate of Economics and Statistics/State Groundwater Department/Central Groundwater Board. Thus, there was a need to develop an index of interference among irrigation wells using the available data. Since interference among irrigation wells is imminent, and since it is difficult to specifically point at well A or well B as the causing agent of interference, it is appropriate to develop an index of *cumulative* interference of irrigation wells rather than interference of irrigation wells. Five indices of cumulative interference were accordingly developed using the available secondary data at the taluk (block) level, and the taluks affected by interference were discussed with hydrogeologists. The indices developed are discussed below.

Index of Cumulative Interference 1 The concentration of the number of irrigation wells or irrigation pump (IP) sets per hectare of net sown area = (no. of irrigation wells or IP sets/net area sown), which is computed for each taluk. The taluks are then arranged in descending order of magnitude reflecting high to low index of cumulative interference.

Implication According to index 1, taluks topping in interference were Honnavar and Kundapura, located in the coastal area. However, these taluks, even though they have high well density, also receive fairly high rainfall and have a relatively better groundwater potential. Thus, this index hides the volume of groundwater availability for irrigation and accordingly is a weak index of cumulative interference among wells.

Index of Cumulative Interference 2 The concentration of a number of irrigation wells or IP sets per million cubic meters (MCM) of utilizable groundwater for irrigation = (number of IP sets/utilizable groundwater for irrigation in MCM) for each taluk.[1] The taluks are then sorted in descending order of magnitude of this index. This index subsumes the effect of draft–recharge ratio since the numerator is a close proxy for draft, and the denominator is a close proxy for recharge. It is recognized that crop pattern is a crucial component of interference. However, at the macro level it is difficult to capture the crop pattern. If a crop pattern is irrigation-intensive, then the higher the index, the higher the irrigation-intensive crop pattern. In Karnataka, around 95 % of the IP sets were of 5 HP capacity and accordingly, index 2 will not be highly sensitive to the horse power (HP) of the IP set.

Implication According to index 2, taluks topping in interference were Devanahalli, Malur, Naragund, Hukkeri, Bangalore North, Kolar, Hosakote, Chikodi Channapattana, Gubbi, Shidlaghatta, Chikkaballapur, Anekal and the like. The taluks which did not show interference problems were Supa, Hosanagara, Narasimha Raja Pura, Mundagod, Bhadravathi, Sagar, Soraba, Jeevaragi, Sringeri, Kundagol, Chitapur, Karwar and others. The hydrogeologists of the State Groundwater Department and the Drought Monitoring Cell of the Government of Karnataka corroborated the results from this index, and several of the taluks were already classified as 'dark' areas.

Thus, index 2 offered a reasonably good estimate of taluks suffering from cumulative interference among wells. The majority of such taluks were located in the eastern dry agroclimatic zone, which has no perennial rivers. In order to locate taluks suffering from interference, this index was developed for taluks located in major dry and transitional agroclimatic zones of Karnataka such as the northeastern transition zone, northern transition zone, northern dry zone, central dry zone, eastern dry zone and southern dry zone. The hilly zone, coastal zone, southern transitional zone and northeastern dry zone were not considered as these zones have a fairly good source of rainfall and, accordingly, groundwater was relatively not as important as in other agroclimatic zones.

Index 2 is computed for all villages in the selected taluk with high interference and then sorted in descending order. Two villages with the top two values of index 2 were chosen as high interference villages, and two villages with the bottom two values of index 2 were chosen as low interference (or control) villages in the high interference taluk. In this process, villages with any kind of surface irrigation facility (from major, medium and minor irrigation sources) were excluded, and only those villages which did not have any surface irrigation facility were chosen. Later, the villages so chosen were surveyed to confirm the prevalence of the cumulative interference phenomenon. After the choice of villages, a participatory rural appraisal (PRA) map was drawn to map the irrigation wells, and 40 farmers owning irrigation wells in that part of the village with a high density of irrigation wells were

[1] The points provided by Sri V. Jagannathan, Senior Hydrogeologist, Central Groundwater Board, Southwestern Region, Bangalore with regard to this index are greatly appreciated.

interviewed. This exercise was repeated in low-interference villages with another sample of 40 farmers, for comparison as control. Since well interference is imminent, farmers in 'high' interference situations are compared with farmers in 'low' interference situations rather than 'no' interference situations.

Index of Cumulative Interference 3 The number of IP sets per MCM of utilizable groundwater divided by gross area irrigated is the index. The taluks identified according to index 3 were not corroborated by the hydrogeological experts, and hence this index was dropped.

Index of Cumulative Interference 4 The proportion of dug-cum-borewells out of total dug wells was another criterion. The taluks identified according to index 3 were not corroborated by the hydrogeological experts, and hence this index was dropped.

Index of Cumulative Interference 5 The groundwater use intensity index (GUII) is the ratio of the number of wells or IP sets per MCM of utilizable groundwater for the taluk in relation to the state and the gross area irrigated by wells per MCM of utilizable groundwater in each taluk in relation to the state figure:

$$\frac{(\text{Number of IP sets/utilizable groundwater in MCM})\ \text{for each taluk}}{(\text{Number of IP sets/utilizable groundwater in MCM})\ \text{for the state}}+\frac{(\text{Gross area irrigated by IP sets/utilizable groundwater in MCM})\ \text{for each taluk}}{(\text{Gross area irrigated by IP sets/utilizable groundwater in MCM})\ \text{for the state}}$$

The taluks were sorted in the descending order of magnitude of index 5 (or GUII).

Implication ndex 5 provided a twin measure of groundwater interference, the number of wells per MCM of utilizable groundwater and the gross area irrigated by wells per MCM of utilizable groundwater. Upon comparison of the results of index 2 and index 5, almost the same set of taluks was identified by these two indices. Accordingly, index 2, which is relatively easier to estimate, was retained.

8.2 Growth in Groundwater Investment

With regard to the investment in groundwater irrigation, since the electricity used to pump out groundwater in India is either provided free of cost or charged on a flat basis, the marginal cost of lifting groundwater becomes modest. However, other explicit and implicit costs of groundwater extraction exist. Currently, around 80 % of the irrigation is met by groundwater in India, and this varies over space and time. Such appreciable irrigation has been possible only with an increase in private investment in irrigation wells by farmers in the HRAs in India. In these areas, the yield of groundwater is dependent, inter alia, on the groundwater recharge effort, the degree of well interference, the crop pattern and the quality of power supply. As the HRAs do not have access to perennial rivers, and as the only recourse is groundwater, farmers are increasing their investment on this fragile and uncertain resource. Groundwater is a natural resource with no clear property rights. Hence, there is market failure. Its occurrence in the unconfined aquifers in the HRAs is sensitive to interactive effects of wells and renders it as a fugitive resource. In the HRAs, about 90 % of the aquifers are unconfined. The nature of groundwater rights is thus intricate.

8.3 Dynamic Nature of Rights

Property rights to groundwater[2] are dynamic and vary with the supply and demand side factors, which jointly determine the property rights. A farmer who is an early comer in groundwater irrigation growing low water-intensive crops, lifting water from a dug well with manual lifts, almost feeling that he/she is enjoying (permanent) private property rights to groundwater will suddenly be shattered once a set of neighbouring farmers emerge who tap groundwater from deeper layers from borewells causing cumulative well interference effects, resulting in permanent failure of the dug well.

8.4 Property Rights to Groundwater Are Obscure

Groundwater is an indispensable resource for irrigation in many pockets of HRAs, especially where there are no flows of perennial rivers. In India, the rights in groundwater belong to the landowner as groundwater is attached to the land property. There is no limitation on the volume of groundwater extraction by a landowner. Since landownership is a prerequisite to the ownership of groundwater,[3] it is difficult to assign an 'open access' nature to groundwater resources. Though landowners own groundwater de jure, this right is limited by the huge investment necessary to tap the groundwater by the construction/drilling of irrigation well(s) and high well failure probability, which allows a selected few among them to have access to groundwater. Unless groundwater is tapped in a well and is available, there is no accessibility since there is no guarantee that any landowner who attempts to construct/drill a well is assured of groundwater, even for a short period. Initial failure and falling life of irrigation wells are a common feature in HRAs. In the eastern dry zone of Karnataka, the (negative binomial) probability of well failure is estimated to be 40 %,[4] which means that a farmer has to drill at least two wells, one of which may be successful. It is crucial to realize that 'wells can exist without groundwater' and not vice versa!

---

[2] The section on the nature of property rights and externality is drawn from Chandrakanth et al. (1997).

[3] Singh (1992).

[4] Nagaraj et al. (1994).

Under these circumstances, the groundwater rights are obscure since farmers are tapping the resource with myopic behaviour, not recognizing the fact that each one's extraction is a function of the neighbouring well's extraction at a time and over time. This, over time, leads to cumulative interference of wells and reduction in life of the wells and in the gross area irrigated by wells. Wantrup[5] indicates that groundwater is a 'fugitive resource' since 'definite property rights belong only to those who are in possession—that is who gets there fastest with the mostest'. *Hence, any estimate of groundwater value (or price or cost) is at the most its shadow price and is subject to addition of different layers of transaction costs involved in each specific aquifer.*

8.5 Intertemporal Externality

In groundwater, the intertemporal externality is the externality imposed by each well-owning farmer on another well-owning farmer over time at a given space. *Due to intertemporal externality, the drilling costs increase as the water table falls.*[6] The externality affects both the poor and rich people, both spatially and temporally and has serious equity implications. Obviously the recently drilled borewell farmers gain better physical and economic access to groundwater when compared with the traditional dug well farmers.

8.6 Policy Governing Groundwater Irrigation

As outlined above, due to obscure property rights to groundwater for irrigation and with a large number of farmers possessing irrigation wells, any policy governing groundwater irrigation can seldom be effective/efficient/equitable/sustainable in letter and spirit. Nevertheless, a thorough groundwater policy is the need of the hour to discipline groundwater users and uses. A crucial requirement of this policy is the creation of a 'well log' at village level, to increase accountability of groundwater users. This needs documenting relevant data for all wells including irrigation wells such as type of well, date of drilling/construction, dimensions of the well, depth of casing, location of the well, how far it is located from the nearest wells in different directions, whether any surface body is available and if so, at what distance and slope, type of IP (surface or submergible), standard of pump (ISI or not), pump HP and number of stages, type of delivery pipe (GI/PVC/HDPE pipe), pumping lift, source of power (diesel or electricity), distance to which groundwater is transported through conveyance pipe (if any), type of irrigation (flow, basin, flood, sprinkler, drip) and the associated discharge of the pump, yield of irrigation well (gallons per hour) at the time of drilling, date energized, number and types of wells constructed/drilled earlier

[5] Ciriacy-Wantrup, S.V. (1968)

[6] Dasgupta (1982).

on the farm with their investment and the number of years for which the irrigation wells have been functioning, investment in irrigation well/s, type of soil, cropping pattern followed in the command area of the irrigation well on the farm including the area under different crops in different seasons. These data are required, in addition to fixing (i) an electrical meter and (ii) a water flow meter to measure the electricity use and for volumetric measurement of groundwater extracted.

Successive governments have been soft towards groundwater extraction for agricultural purposes as it has turned out to be a political economy question. In Karnataka, the State Electricity Board has always overestimated the use of electricity for agriculture, by treating it as the residual, after providing for all uses, and even accounting a sizeable portion of transmission and distribution (T&D) losses for agriculture. Other states may also have a similar pattern of estimating power use in agriculture in the absence of electrical meters for IP sets.

In one of our studies, it was estimated that, to lift one acre-inch of groundwater for irrigation from a depth of around 200 ft, the IP set uses 42 kW h of electricity. However, this cannot be linearly extrapolated over increasing depth. Thus, even if the electricity is priced at ₹ 1 per kilowatt hour, it costs ₹ 42 per acre-inch of groundwater. Estimates of the State Electricity Board highlight that a typical irrigation well uses around 6000 kW h or units per year. A typical borewell with 500 ft depth yielding around 1500 gallons/h, using 6 h of electricity per day for about 300 days in a year, pumps out around 120 acre-inches of groundwater. With 6000 kW h of electricity charged at around ₹ 1 per unit, this works to around ₹ 50 per acre-inch. Thus, the estimate of 42 kW h, though crude, cannot be discounted. But what is more important is that the value or cost of groundwater could be much more than the cost of electricity to pump out, due to the cumulative interference of irrigation wells. Currently, according to the latest National Sample Survey Organization study, in India, around 66 % of irrigation wells use diesel for running the pumpsets, while only 34 % use electricity. Hence, if free or flat-rate electricity is the prime cause for overextraction, then the farmers would not have switched to using diesel in such appreciable proportions. Thus, the policy of providing electricity at zero marginal cost cannot solely be responsible for the mushrooming of wells. Other factors, inter alia relative prices of crops including support price for different crops, soft loans for drilling/constructing irrigation wells, increasing effective demand for vegetables and flowers in periurban areas, expanding access road network and thus linking markets effectively, have also contributed towards the secular overdraft of groundwater in already overexploited areas.

8.7 Relevance of Valuing Groundwater

Farmers in general do not pay for the electricity used in pumping groundwater for irrigation in most parts of India. Thus, the marginal cost of extraction of groundwater is modest, as there are no operational costs of payments towards the use of electricity other than repairs and maintenance. It is crucial to note that electricity is not the only crucial cost or implicit benefit (since there is a subsidy) to groundwater farmers. The farmers are in fact incurring much more on groundwater due to the costs imposed by negative externality resulting from cumulative interference of irrigation wells. Thus, in the context of natural resource economics, valuing groundwater for irrigation is relevant for farmers in HRAs due to the high probability of well failure which forces the farmers to invest in additional well(s). The forced investment due to initial and premature failure of wells is considered a negative externality. These negative externalities are implicitly incurred through forced investment on, inter alia, (i) additional well(s), (ii) deepening of existing wells since the well(s) constructed earlier failed to yield water for the expected number of years and/or (iii) water storage structures (such as overground storage structures, conveyance pipes, drip irrigation, sprinkler irrigation, etc.) in order to at least remain on the original isorevenue curve. Cumulative interference among irrigation wells is the major factor responsible for both reciprocal and intertemporal externality in using groundwater for irrigation in HRAs. We attempt basically to value or cost groundwater with this major hypothesis.

The annual cost of irrigation = amortized cost of irrigation well + amortized cost of conveyance + amortized cost of pumpset and electrical installation + annual cost of repairs and maintenance. The labour cost of irrigation is merged with the cost of labour for other cultural operations. Thus, the labour cost involved in irrigation is not considered in the cost of irrigation. The amortized cost of the well includes amortization of investment on all the wells on the farm (working and failed wells). The amortized cost of irrigation of each irrigation well on the farm is added across all wells on the farm. This total amortized cost of irrigation is then apportioned over the total groundwater extracted in order to obtain the amortized cost per acre-inch of groundwater. The cost of cultivation is obtained as the summation of the cost of human labour, bullock labour, seeds, fertilizers, plant protection measures, application of silt/manure, transportation and bagging, the amortized cost of irrigation and the opportunity cost of working capital.

Cost of production is cost of cultivation + amortized cost of irrigation + interest on variable cost + opportunity cost of dry land agriculture. The opportunity cost of dry land agriculture is the average net return obtainable from that piece of land devoted to irrigation if that land was cultivated on rainfed basis. Gross returns for each crop is the value of the output at the prices realized by farmers. Net returns from well irrigation is the gross return from gross irrigated area minus the cost of production of all crops.

8.8 Estimation of the Cost of Electrical Energy Used

An electrical meter needs to be fixed to measure the electricity used for irrigation. For demonstration purpose, an electrical meter was fixed to a submersible irrigation borewell of 6 in diameter, and 5 HP capacity lifting groundwater in Tiptur area (central dry agroclimatic zone) from a depth of 200 ft, to support 7 acres of coconut plantation and 2 acres of paddy crop.[7] Every time the pump was put on and

[7] Satisha (1997).

put off, the voltage, the initial reading and the final reading were noted. In 1 year, 4368 kW h of energy were used to lift 104 acre-inches of groundwater (1 acre-inch = 22,611 gallons of water). If electricity is valued at the rate of ₹ 1 per kilowatt hour, (50 paisa higher than what is recommended by the Central Electrical Authority of India), then the total cost of electricity is ₹ 4368 or ₹ 42 per acre-inch of groundwater. The electricity cost to lift groundwater varies with the pumping lift, type, HP, quality and age of pump, frequency and depth of irrigation as decided by the crop pattern and soil type and method of irrigation. The measurement of electricity cost to lift groundwater necessarily requires the cooperation of the farmer for whose well the electrical meter is fixed. For this study, even though electrical meters were procured in ample numbers, only one farmer cooperated by agreeing to the installation of an electrical meter as they were apprehensive of being charged for the electricity. Thus, in this study, the electricity cost estimated in the central dry agroclimatic zone is used for all zones, in the absence of such estimates from other zones. Hence, the electricity cost is invariant over zones and across different degrees of interference.

8.9 Estimated Groundwater Values

The estimated value per acre-inch of groundwater (excluding the cost of electricity) (Table 8.1) varied from ₹ 74 per acre-inch in the southern dry zone to ₹ 2700 in the southern transitional zone considering that these results pertain to years before 2000. Obviously the value (cost) of groundwater would increase with increase in probability of premature and initial well failure, depth of drilling, number of wells on the farm, externality faced by farmers and other variables already discussed. Such a wide variation in the value of groundwater is due to the varying hydrogeological conditions, crop pattern and level of interference. The irrigation cost as a percentage of the total cost varied from 22 % in the eastern dry zone to 70 % in the southern transition zone. The value of groundwater forms a substantial proportion of the total cost of the production of crops. The huge cost of groundwater irrigation is compensated by the product price of most of the crops supported by groundwater in the HRAs. However, this situation may not hold good all the time due to dynamic market forces.

8.10 Electricity Subsidy

Considering the different values of groundwater (Table 8.2), an addition of ₹ 42 per acre-inch to the value of groundwater brings down the net returns of farmers marginally since the major component of the value of groundwater is dominated by the implicit costs of well failure reflected in the negative externality. Charging for electricity influences the farmers in terms of using scarce groundwater to cultivate water-intensive commercial crops when compared with water-intensive crops like paddy and sugarcane, wherever water use intensity is high. In other areas too,

Table 8.1 Net return considering value of groundwater and cost of electrical energy across agroclimatic zones of Karnataka

Zone	Well interference	Gross Irrigated Area (GIA) per farm (acres)	Dominant crops	Water used per farm (acre-inches)	Method of irrigation	Water used per acre of GIA (acre-inches)	Estimated kW h of electricity used to lift ground water for 1 acre of GIA	Value (cost) per acre-inch of ground-water	Cost of electric-ity to lift ground-water for 1 acre of GIA @ ₹ 1 per kW h	Net returns per acre of GIA (after including value of ground-water and cost of electricity)	Net returns per acre-inch of water (after including value of ground-water and cost of electricity)
Southern Transi-tional Zone—Channagiri taluk	HIGH	5.16	Arecanut, cotton, onion	86.6	Drip,	16.78	705	2700	705	23912	1425
	LOW	5.04		95.2	Flood	18.9	794	529	794	12166	644
Central Dry Zone—Madhugiri taluk	HIGH	5.32	Paddy, arecanut	37.4	Flood	7.03	295	457	295	167	24
	LOW	5.74		59.8		10.42	438	211	438	303	29
South-ern Dry Zone—Chama-raja-nagar taluk	HIGH	8.58	Sugarcane, paddy	134	Flood	15.6	655	104	655	2273	146
	LOW	11.32		145		12.8	538	74	538	2069	162
North-ern Dry Zone—Athani taluk	HIGH	2.99	Local maize, jowar, groundnut, turmeric	4.6	Flood	1.54	65	972	65	3354	2178
	LOW	3.76		10.9		2.90	122	480	122	1978	682

Table 8.1 (continued)

Zone	Well interference	Gross Irrigated Area (GIA) per farm (acres)	Dominant crops	Water used per farm (acre-inches)	Method of irrigation	Water used per acre of GIA (acre-inches)	Estimated kW h of electricity used to lift ground water for 1 acre of GIA	Value (cost) per acre-inch of ground-water	Cost of electricity to lift ground-water for 1 acre of GIA @ ₹ 1 per kW h	Net returns per acre of GIA (after including value of ground-water and cost of electricity)	Net returns per acre-inch of water (after including value of ground-water and cost of electricity)
Eastern Dry Zone—Malur taluk	HIGH	3.85	Tomato, potato, rose	57.4	Flood	14.9	626	287	626	8024	539
	LOW	3.50	Ragi, tomato, mulberry	37.8	Flood	10.8	454	394	454	6059	561
Eastern Dry Zone—Devanalli taluk	HIGH	7.83	Mulberry, grapes, ragi	45	Flood	5.70	239	308	239	3772	662
	LOW	5.43	Ragi, potato, mulberry	81	Flood	14.90	626	200	626	1775	119

Table 8.2 Average electricity used and net returns realized by groundwater farmers in Karnataka

Level of well interference	Average electricity used to lift groundwater for 1 acre of gross irrigated area (kW h)	Coefficient of variation (%)	Average net returns per acre of gross irrigated area (₹)	Coefficient of variation (%)
High	431	62	8715	126
Low	495	45	4412	109

pricing electricity has the potential to influence the crop pattern. The weighted average price of electricity will vary, inter alia, with the type of well, depth of well, type and age of pump, suction and delivery heads, type and quality of suction and delivery pipes, level of groundwater use, number of IP sets on the farm, supply of electricity at the right voltage, number of water storage structures on the farm and the use of pump to lift that water, method of irrigation followed (flood, sprinkler, drip irrigation), the type of crop and soil. In some pockets, farmers have even resorted to the use of diesel generators to energize pumps to lift groundwater due to vagaries of electricity supply. The question of whether pricing electricity for lifting groundwater would facilitate sustainable use of groundwater depends on the market forces for crops.

In a study in Tamil Nadu,[8] it was found that even if the power tariff was increased by 100 %, the present value of net benefits fell only by 8 %, that too under optimal control. Even considering extraction under myopic conditions, where overexploitation is a rule, the fall in present value of net benefits was virtually 0 %, for a 100 % increase in the price of electricity. According to this study, electricity pricing neither reduced the extraction of groundwater nor did it bring substantial losses to farmers. Another study conducted by the Tata Energy Research Institute, New Delhi in 1996 arrived at similar conclusions. Pricing electricity is a sensitive issue. However, if farmers can be convinced about the depleting groundwater levels and the need to bring wise use norms, it would result in disciplining farmers towards judicious use of groundwater.

8.11 Groundwater Markets Uncommon in Peninsular India

Markets can exist if there is marketable surplus. Groundwater markets are not a common phenomenon in peninsular India due to physical and economic scarcity of groundwater. If electrical subsidy were to be a windfall gain to farmers in this region, and if groundwater were to be available, groundwater markets would flourish. Groundwater rents are increasing due to increasing probability of well failure and the interactive effects of wells. Improvement in groundwater recharge on the supply side and institutions to discipline groundwater use on the demand side can shape a wise groundwater use phenomenon in Karnataka. It has been found that in

[8] Balasubramanian (1998).

areas where the groundwater recharge induced through desiltation of the irrigation tank is effective, the groundwater rent reduced by 27 % (in Malur taluk in the eastern dry zone).

8.12 Value of Groundwater in Cost of Production

Valuing groundwater for irrigation provides a yard stock of the level of contribution of groundwater to irrigation. The value of groundwater used forms a substantial portion of the total cost of production in the HRAs of peninsular India. The cost of electricity to lift groundwater results in a modest rise in the value (cost) of water and forms a modest proportion of the cost of cultivation. It varies across different zones, different degrees of interference and variation in groundwater applied per acre. A major proportion of groundwater rent is the magnitude of negative externality involved in groundwater resource use. For reasons of fiscal probity and to generate information regarding actual use of groundwater, it is essential to meter the electricity use as well as the groundwater use and charge progressively on a slab basis by educating farmers and other actors. In order to optimally exploit groundwater, policies concerning groundwater should consider the role of surface water bodies like irrigation tanks, canals, reservoirs and watersheds. Workable programs and policies should be specific to each condition.

Policies of isolation distance, credit supply, power pricing and subsidies to adopt water saving technologies are sine qua non for conservation of groundwater resources. There cannot be a uniform isolation distance norm since the proximity to surface water bodies and groundwater recharge capacity can mask the effect of the high density of irrigation wells. The cropping pattern plays a significant role in groundwater use and conservation. Irrigation literacy should be promoted among farmers towards sustainable use of the economically scarce groundwater.

8.13 Variation in Net Returns and Electricity Used

The average level of use of electricity to lift groundwater to irrigate 1 acre of gross irrigated area varies between 431 kW h in the high well interference area and 495 kW h in low interference situations (Table 8.2). The variation is between 62 and 45 % for these situations. However, considering the average net returns per acre, the net return is ₹ 8715 per acre of gross irrigated area in high interference areas, while in low interference situations the net return is ₹ 4412 per acre. This indicates that the crop pattern which is decided by the effective demand for different commodities will have a strong influence on the net returns. In addition, the wide variation in coefficient of variation exceeding 100 % in both the areas indicates the problems in averaging the land requirement for a desirable level of living or profitability in different areas of the state concerning the use of groundwater.

Since the level of groundwater extraction is about 30 % of the potential in Karnataka, and even assuming that there is low well interference in the regions which hold 70 % of the usable potential of groundwater, the huge variation of 109 % in the average net returns of ₹ 4412 per acre of gross irrigated area reflects the complexities involved in advocating the level of net return which enables farmers to make a comfortable living. However, these give an indication of the potential net return which can be generated even by accounting negative externalities and paying for electricity to lift groundwater.

8.14 Epilogue

Considering the five agroclimatic zones in Karnataka, the average gross irrigated area per groundwater-irrigated farm is 5.71 acres, applying 11.6 acre-inches of groundwater per acre of land, using 487 kW h of electricity to lift groundwater for 1 acre of gross irrigated area. With an estimated use of 42 kW h to lift 1 acre-inch (22,611 gallons) of groundwater, at a charge of ₹ 1 per kilowatt hour to the agriculture, the total electricity bill amounts to ₹ 2781 per annum for using a total of 66.24 acre-inches of groundwater. If the electricity is charged at a flat fee of ₹ 300 per HP, assuming that an average groundwater-irrigated farm has only one IP set of 5 HP capacity, the annual electricity payment is ₹ 1500. Hence, the flat fee of ₹ 300 per HP can recover 54 % of the estimated electricity used by irrigated farms if 1 kW h is charged at ₹ 1. Obviously, if 1 kW h for agriculture is charged at 50 paisa, the estimated electricity bill is ₹ 1390, and the existing rate of ₹ 300 per HP would suffice to recover the electricity charges, including a part of other contingent expenditure. The net returns realized (after accounting for value of groundwater and electricity charges) work out to ₹ 5488 per acre of land totalling to ₹ 31,336 for a groundwater-irrigated farm per year.

Every year, 360.96 km^3 of groundwater resources are available for irrigation in India, of which only 115.17 km^3forming 31.91 % is extracted (as net draft). For Karnataka too, the situation is no different. Karnataka has 13.76 km^3 of groundwater resource, of which 4.3 km^3 is extracted (as net draft) forming 31.26 % (Source: Water and related statistics, Central Water Commission, Government of India, 2000, Table 1.26, p. 51). However farmers need to be educated that unless there is adequate effort to recharge groundwater, they cannot go on pumping groundwater in the same manner as they are used to. Many farmers in Kolar district for instance, have voluntarily switched to drip irrigation even for close-spaced crops such as vegetables. This shows that it is not electricity or energy to lift water alone, but it is groundwater which is crucial and vital for farmers. The culture of wise use of groundwater has to be inculcated and appreciated by farmers. Thus, they need to be educated regarding water literacy. In order to promote the education of farmers regarding wise use of groundwater, flow meters (water meters) need to be installed onto IPs and demonstrations need to be held for farmers regarding the volume of groundwater extracted and used every time they put on the pump. In addition, if there is a well in the proximity belonging to the farmer or his/her neighbour, he/she can also see the mutual interference among wells. This also helps the farmer to budget his water for different crops. Finally, this helps him to appreciate paying (in part or full) for electricity. Without the cooperation of farmers, no groundwater regulation can be effective in India. Thus, merely passing a groundwater law is no panacea without irrigation literacy. Parallel to agricultural extension, unless there is irrigation extension, farmers would seldom appreciate the value of the water they are using. A crucial point to be noted is that cumulative interference is becoming a predicament due to increasing groundwater development in areas where groundwater is already being extracted. In addition, watershed development programs and water recharge efforts and incentives for both are crucial in India, which augment the crucial groundwater resource for irrigation.

The state of Gujarat has contributed richly to literature on economics of groundwater through its illustrious son, Professor Tushar Shah. The state is emerging as a role model for economic development with commitment and good governance in all spheres. The state can improve its services to farmers by launching a separate Department of Irrigation Extension to be the pioneer in educating farmers regarding, inter alia, agronomic, economic, hydrologic and hydrogeological aspects of groundwater and surface water for irrigation, which will pave the way for realization of economic value of water by farmers, who use 85 % of all water resources for irrigation.

References

Balasubramanian R (1998) An enquiry into the nature, causes and consequences of growth of well irrigation in a vanguard agrarian economy, Unpublished Ph.D thesis, Department of Agricultural Economics, Tamil Nadu Agricultural University, Coimbatore, India

Chandrakanth MG, Arun V (1997) Externalities in groundwater irrigation in hard rock areas. Indian J Agric Econ 52(4):761–771

Ciriacy-Wantrup SV (1968) Resource conservation: economics and policies. Division of Agricultural Economics. University of California, Berkeley, pp 142

Dasgupta Partha (1982) The control of resources. Harvard University Press, Cambridge

Nagaraj N, Chandrakanth MG, Gurumurthy (1994) "Borewell failure in drought-prone areas of Southern India: a case study". Indian J Agric Econ 49(1):101–106

Satisha KM (1997) Resource economics study of valuation of well interference externalities in Central Dry Zone of Karnataka, Unpublished MSc (Agri) thesis submitted to the Department of Agricultural Economics, University of Agricultural Sciences, Bangalore

Singh Chatrapati (Ed) (1992) Water Law in India. The Indian Law Institute, New Delhi

Chapter 9
Demand Side Economics of Micro-irrigation

9.1 Preamble

The first experimental system of drip irrigation was established in 1959 by Netafim, an irrigation company by Blass in partnership with Kibbutz Hatzerim in Israel. They developed and patented the first practical surface drip irrigation emitter. In the USA, the first drip tape, called *Dew Hose,* was developed by Richard Chapin in the 1960s. In India, the Jain Irrigation Company heralded drip (micro)-irrigation in 1989 developing the 'integrated system approach'.

Of 10.80 million ha of cropped area in Karnataka, 21.50 % is irrigated and 78.5 % is rain fed. Two thirds of the geographical area is in the semi-arid zone receiving less than 750 mm of rainfall and suffering frequent droughts. Progressive farmers in dry land began using the drip system in the late 1970s without any support from the state. Later, due to policy support, use of the drip irrigation system (DIS) spread primarily to irrigate high-value dry land horticultural crops (Shashidhara et al. 2007). Maharashtra witnessed a steep rise in the use of DIS in 1988. Currently, the total area under drip irrigation is 0.62 ml ha in India and 0.114 ml ha in Karnataka.

9.1.1 Status

Karnataka ranks third in the installation of drip irrigation in India. In recognition of the increasing probability of well failure to the tune of 0.4 in Karnataka, the state provided a subsidy of 50–100 % for drip installation through the departments of horticulture and agriculture. The subsidy amount accounted for ₹ 95.53 million, covering an area of 8284 ha. Among horticultural crops, coconut ranks the first (27,784 ha) in area, followed by arecanut (11,139 ha), mango (6286 ha), grapes (3983 ha), sapota (1139 ha) and flowers and vegetables in the area covered by drip irrigation in Karnataka. Under the National Horticultural Mission, farmers in 25 of the 27 districts in Karnataka will get a 75 % subsidy for installing drip irrigation equipments, while those in Bijapur and Kolar districts will get a 100 % subsidy.

M.G. Chandrakanth, *Water Resource Economics,* DOI 10.1007/978-81-322-2479-2_9

Karnataka, the second dry region in the country after Rajasthan, is a pioneer in implementing drip and sprinkler irrigation to save water, power and labour and also help farmers to cope with economic scarcity of groundwater. Of the 16.30 lakh hectares of land, only 1.64 lakh hectares have been drip irrigated.

In Karnataka, drip irrigation is common in the eastern dry agroclimatic zone of Karnataka due to acute economic scarcity of groundwater (Chandrakanth and Romm 1990). Here grapes, mulberry, tomatoes and other vegetables are popularly cultivated. A sample of 90 farmers consisting of 45 farmers using groundwater with drip irrigation and another 45 farmers using groundwater with conventional irrigation growing grapes, mulberry and tomatoes on their farms in Chickballapur district were chosen for this study (Priyanka 2009). Primary data for the study was obtained from personal interviews during Dec 2008–Jan 2009, using a structured pre-tested schedule. The information elicited included inter alia (1) socioeconomic features of respondents, (2) cropping pattern (3), land holdings (4), sources of irrigation, (5) investment on irrigation wells (6), costs and returns from crops grown under well irrigation and (7) volume of water used measured in acre-inches (1 ac-in. = 22,611 gallons of water = 1 ha cm).

Economic efficiency of water use was analysed in terms of net income per acre-inch of water and technical efficiency in terms of output per acre-inch of water used. Regression analysis was used to find out the factors influencing the net returns per farm (Chandrakanth and Arun 1997). The Tobit model was used to estimate a farmer's willingness to invest in a DIS and discriminant analysis used to find the variables that differentiate drip irrigation farmers (DIF) from conventional irritation farmers (CIF). The results are largely drawn from Chandrakanth et al. (2013).

9.1.2 Cropping Pattern

Considering gross cropped area (GCA), in drip farms the area under mulberry (sericulture) was around 40 %, the area under grapes was around 48 %, ragi was 4.35 % of the GCA in kharif and 3.72 % in rabi. In conventional farms, the area under mulberry was 33 %, grapes was 8 %, ragi was 18 % in kharif, 9 % in rabi and in summer, was 5 % of the GCA. The cropping intensity was comparable in both situations (Table 9.1). The area under perennial crops enabled farmers to adopt drip irrigation compared with annual crops dominating in conventional irrigation farms. However, whether drip irrigation was introduced first or the perennial crops were introduced first is a hen–egg question. In order to cope with water shortage for the already planted perennial crops, drip irrigation was introduced.

9.1.3 Well Failure

In drip irrigation farms, 36 % belonged to marginal farms with an average size of holding of 1.39 ac, 51 % of the farms belonged to small farms with an average holding size of 3.55 ac and 13 % belonged to large farms with a holding size of

Table 9.1 Cropping pattern in drip and conventional irrigation farms in the eastern dry zone of Karnataka (acres)

Crops	Drip irrigation farms ($n=45$)		Conventional irrigation farms ($n=45$)	
	Area (acres)	Proportion of GCA	Area (acres)	Proportion of GCA
Kharif				
Ragi	12.00	4.35	44.00	18.24
Jowar	3.00	1.09	7.00	2.90
Horse gram	–	–	3.50	1.45
Beans	–	–	1.50	0.62
Tomato	3.00	1.09	14.00	5.80
Potato	–	–	5.50	2.28
Chili	–	–	2.00	0.83
Subtotal	18.00	6.53	77.50	32.12
Rabi				
Ragi	10.25	3.72	23.00	9.53
Jowar	–	–	2.00	0.83
Tomato	–	–	5.00	2.07
Potato	–	–	2.00	0.83
Chili	–	–	4.00	1.66
Subtotal	10.25	3.72	36.00	14.92
Summer				
Ragi	3.50	1.27	11.25	4.66
Tomato	1.00	0.36	17.00	7.05
Subtotal	4.50	1.63	28.25	11.71
Perennials				
Grapes	133.50	48.41	20	8.29
Mulberry	109.50	39.71	79.50	32.95
Subtotal	243.00	88.12	99.50	41.24
Gross cropped area	275.75	100	241.25	100.00
Net cropped area	139.50	–	127.25	–
Cropping intensity	197.67	–	189.58	–

GCA gross cropped area under grapes is considered as twice their net area, and mulberry is considered as two times of the net area giving weightage to the perennial crops

8.75 ac. Considering all the farms, 33 % of wells had failed, and 67 % of wells were functional at the time of data collection (2008). The earliest well was drilled in 1958 and the latest during 2007. In conventional irrigation sample farms, 38 % farms belonged to marginal farms with an average size of holding of 1.17 ac, 60 % of the farms belonged to small farms with an average holding size of 3 32 ac, and 2 % belonged to large farms with a holding size of 6 ac. Considering all the farms, 19 % of the wells had failed, and 81 % were functional at the time of data collection (2008). The earliest well was constructed in 1958, and the latest was drilled during 2008. (Table 9.2).

Table 9.2 Distribution of irrigation wells across the size of holding in the eastern dry agroclimatic zone (EDZ) of Karnataka

Type of farms	Size of holding (acres)	No. of farms (%)	Functioning/working wells (%)	Non functioning/failed wells (%)	Total no. of wells	Range of years of drilling
Drip irrigation farms						
Marginal farms (<2.5 ac)	1.39	16 (35.56)	16 (61.53)	10 (38.46)	26	1970–2006
Small farms (2.5–5 ac)	3.55	23 (51.11)	35 (71.43)	14 (28.57)	49	1958–2007
Large farms (>5 ac)	8.75	6 (13.33)	13 (65.00)	7 (35.00)	20	1976–2006
All farms	3.48	45	64 (67.37)	31 (32.63)	95	1958–2007
Conventional irrigation farms						
Marginal farms (<2.5 ac)	1.17	17 (37.78)	17 (80.95)	4 (19.05)	21	1958–2006
Small farms (2.5–5 ac)	3.32	27 (60.00)	28 (80.00)	7 (20.00)	35	1977–2008
Large farms (>5 ac)	6	1 (2.22)	1 (100)	0 (0.00)	1	2004
All farms	2.77	45	46 (80.70)	11 (19.30)	57	1958–2008

Figures in parentheses indicate the percentage to the respective total

9.1.4 Profile of Irrigation Wells

In DIF, 64 (67 %) borewells were functioning, whereas in CIF 46 (81 %) borewells were functioning. Even though the number of wells possessed by farms in DIF was higher as compared to CIF, the proportion of functioning wells was lower in DIF (67 %) as compared to CIF (81 %), and the proportion of well failure was the highest in DIF (33 %) as compared to CIF (19 %).

Considering the investment on irrigation borewells in the two situations, investment per well in DIF (₹ 111,982) was comparable to CIF (₹ 110,165). Investment per functioning well was 26 % higher for farms with drip irrigation (₹ 166,223) compared to conventional irrigation farms (₹ 131,551) because of high well failure.

The amortized cost per well was lower (₹ 17,350) for drip irrigation farms than conventional irrigation farms (₹ 19,196) due to a higher number of wells in DIF (95) than CIF (57). The amortized cost per functioning well was higher (₹ 25,754) for drip farms than CIF (₹ 23,786). The DIF had a higher proportion of well failure (33 %) compared to conventional irrigation farms (19 %). Hence, even though the modal age of wells in both situations are the same, as the well failure and the investment per well was higher due to drip irrigation on DIF, the annual negative externality cost is 85 % higher on DIF compared to CIF (Table 9.3).

Thus, if the amortized cost per well (considering all the wells) is the same as the amortized cost of functioning wells, then there are only functional wells and no failures. If the failure rate is large, then the gap between the amortized cost per well and that per functioning well would also be large, and hence the cost of well failure can be considered as an externality cost (Nagaraj et al. 1994).

9.1.5 Net Returns in Drip-Irrigated Farms and Conventional-Irrigated Farms

The drip farms realized higher net returns from tomato (₹ 26,208) than in farms with conventional irrigation (₹ 22,796). Similarly, in mulberry cultivation, drip farms realized higher net returns per acre per crop (₹ 7621) compared to conventional irrigation farms (₹ 4978). Considering the net returns per acre per crop from grapes, drip farms realized higher returns (₹ 52,084) than that of conventional irrigation farms (₹ 21,489). The net return per acre-inch of groundwater used was higher for drip farms from tomato, mulberry and grapes (₹ 2696, 1384 and 4723, respectively) than for farms with conventional irrigation (₹ 1040, 525 and 769, respectively; Table 9.4).

The net return per rupee of water from tomato, mulberry and grapes realized by drip farms was higher (₹ 2.47, 2.88 and 12.84, respectively) than that by conventional irrigation farms (₹ 2.21, 1.24 and 10.26, respectively). Considering the output per acre-inch of groundwater from tomato, mulberry and grapes, drip farms produced higher (1245, 905 and 1050 kg, respectively) than conventional irrigation farms per crop (503, 386 and 285 kg, respectively; Table 9.4).

Table 9.3 Profile of irrigation wells on drip and conventional irrigation farms

Sl No.	Particulars	Drip irrigation farms	Conventional irrigation farms	% change (drip over conventional)
1	Sample farms (no.)	45	45	–
2	Functioning borewells (no.)	64 (67)	46 (81)	39
3	Non-functioning borewells (no.)	31 (33)	11 (19)	182
4	Total borewells (no.)	95	57	67
5	Average age of functioning wells (years)	9.83	9.28	6
6	Average life of premature failed wells (years)	9.13	7.09	29
7	Average age of all wells (years)	9.58	8.93	7
8	Modal age of wells (years)	9	9	–
9	Depth of borewells (feet)	536	570	−6
10	Yield of well (gallons per hour—GPH)	1663	1739	−4
11	Range of wells drilled (years)	1958–2007	1958–2008	–
12	Earliest year of drip irrigation system installed	2000	NA	–
13	Modal year of drip irrigation	2004	NA	–
14	Investment per well (₹) in 2008	111,982	110,165	2
15	Investment per functioning well (₹) in 2008	166,223	131,551	26
16	Amortized cost per well (₹) in 2008	17,350	19,196	−10
17	Amortized cost per functioning well(₹)	25,754	23,786	8
18	Annual negative externality cost (₹) (17 − 16)	8404	4590	85

Figures in parentheses are the percentage to the respective total *NA* not applicable

Table 9.4 Net returns of crops under drip and conventional irrigation

Per crop of		Potato	Chilli	Tomato	Mulberry[a]	Grapes
Cost of cultivation per acre (₹)	DI	NC	NC	19542	15165	22961
	CI	24471	10520	20793	16939	25686
Gross return per acre (₹)	DI	NC	NC	45750	22786	75045
	CI	36267	18593	43589	21917	47175
Net return per acre (₹)	DI	NC	NC	26208	7621	52084
	CI	11796	8073	22796	4978	21489
Water use per acre (acre-inch)	DI	NC	NC	9.72	5.51	11.03
	CI	22.44	15.24	21.90	9.48	27.94
Net return per acre-inch of water (₹)	DI	NC	NC	2696	1384	4723
	CI	526	530	1040	525	769
Net return per rupee of water (₹)	DI	NC	NC	2.47	2.88	12.84
	CI	1.55	1.95	2.21	1.24	10.26
Net return per kg of output (₹)	DI	NC	NC	2.17	1.53	4.50
	CI	1.50	1.47	2.07	1.35	2.69
Output per acre-inch of water	DI	NC	NC	1245	905	1050
	CI	349	360	503	386	285
Cost per kg of output (₹)	DI	NC	NC	1.61	3.04	1.98
	CI	3.13	3.39	1.88	4.63	3.22
Output per acre (kg)	DI	NC	NC	12100	4981	11577
	CI	7826	5483	11015	3662	7975

NC not cultivated, *CI* conventional irrigation farms, *DI* drip irrigation farms

[a] For mulberry farms, output and returns reflect those from the mulberry leaves excluding cocoons

The net return per acre per crop from mulberry and grapes was higher for drip farms (₹ 7621 and ₹ 52,084) than for conventional irrigation farms (₹ 4978 and 21,489), respectively. Similarly, the net return per acre-inch of groundwater used was higher in drip farms from tomato cultivation (₹ 26,208) than in conventional irrigation farms (₹ 22,796).

9.1.6 *Technical Efficiency*

The technical efficiency of water use is defined in terms of output per acre-inch of water and water used per quintal of output. In the DIF, 9.05 qtl of mulberry leaves, 10.50 qtl of grapes and 12.45 qtl of tomato were produced per acre-inch of water, while in conventional irrigation systems 3.86 qtl of mulberry, 2.85 qtl of grapes and 5.03 qtl of tomato were produced per inch of water. The volume of water used to produce 1 qtl of mulberry, grapes and tomato was lower in DIF (0.11, 0.09 and 0.08 ac-in., respectively) than in CIF (0.26, 0.35 and 0.19 ac-in., respectively). Output per acre-inch of water used in DIF was 234 % higher in mulberry, 368 % higher in grapes and 248 % higher in tomato as compared to CIF. In DIF, 42, 26 and 42 % of water had been saved as compared to CIF in mulberry, grapes and tomato, respectively (Table 9.5).

9.1.7 *Economic Efficiency of Water Use*

The economic efficiency of water use can be defined in terms of net return per acre-inch of water, net return per acre and net return per rupee of water used for irrigation. The net returns per acre-inch of water from mulberry, grapes and tomato were higher in DIF (₹ 1384, 4723 and 2696, respectively) than in CIF (₹ 525, 769 and 1040, respectively). The net return per acre of mulberry, grapes and tomato in DIF was relatively higher (₹ 7621, 52,084 and 26,208, respectively) than in CIF (₹ 4978, 21,489 and 22,796, respectively; Table 9.5). Thus, the technical and economic efficiencies indicate that the positive externality due to groundwater use in drip irrigation over conventional irrigation is higher than the negative externality due to groundwater use in drip irrigation (given by the annual negative externality cost due to well failure).

9.2 Marginal Productivity of Groundwater in Drip and Conventional Irrigation Farms

The net return function per farm was estimated to capture the influence of (i) the volume of water used in drip and conventional irrigation, (ii) dummy variable assigning 0 for conventional farm and 1 for drip farm and (iii) the interaction between

Table 9.5 Water use efficiency in drip and conventional irrigation farms in the eastern dry zone of Karnataka

Particulars	Technical efficiency		Economic efficiency		
	Output per acre-inch of water (quintals)	Water used per quintal of output (acre-inch)	Net return per acre-inch of water (₹)	Net return per acre (₹)	Net return per rupee of water (₹)
Mulberry					
Drip irrigation farms (DI)	9.05	0.11	1384	7621	2.88
Conventional irrigation farms (CI)	3.86	0.26	525	4978	1.24
Efficiency (%)	234	42	264	153	232
Grapes					
Drip irrigation farms	10.50	0.09	4723	52,084	12.84
Conventional irrigation farms	2.85	0.35	769	21,489	10.26
Efficiency (%)	368	26	614	242	125
Tomato					
Drip irrigation farms	12.45	0.08	2696	26,208	2.47
Conventional irrigation farms	5.03	0.19	1040	22,796	2.21
Efficiency (%)	248	42	259	115	112

Efficiency = [DI/CI] × 100

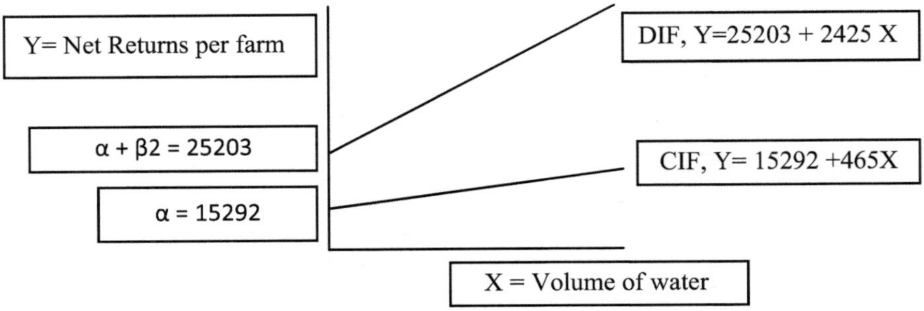

Fig. 9.1 Marginal productivity of groundwater influenced by technology (of drip irrigation). *DIF* drip irrigation farmers, *CIF* conventional irrigation farmers

the method of irrigation and volume of water used (interaction dummy). The dummy variable is used to differentiate the type of irrigation system assigning 1 for drip irrigation and 0 for conventional irrigation farms. The estimated model is $Y=\alpha+\beta_1 X+\beta_2 D+\beta_3 DX+U_i$, where Y=annual net returns obtained in rupees per farm, X=annual irrigation water applied to crops in acre-inches per farm, D is the (0,1) intercept dummy variable representing the shift in the net returns on farms with drip irrigation, DX is the slope dummy variable measuring the rate of increase in net returns due to the interaction of the groundwater volume applied and the drip irrigation method. The resulting net return function with t values in parentheses is depicted in Fig. 9.1.

$$Y = 15292 + 465X + 9911D + 1960\ \mathrm{DX} \tag{9.1}$$

t = (1.41) (2.45) (0.72) (6.17) are the calculated 't' values, Adj $R^2=0.56$, $R^2=0.76^{**}$, F=36, n=81 farmers. ** refers to statistical significance of goodness of fit using F value.

From the regression (1), the threshold net return is ₹ 15,292 per farm which is the contribution of inputs other than irrigation water. The marginal productivity (MP) of groundwater is ₹ 465 per acre-inch at any level of use of water (as this is a linear function) obtained by differentiating the dependent variable with respect to groundwater. Due to drip irrigation, the threshold net return per farm is shifted by ₹ 9911 as given by the coefficient of the dummy variable used to differentiate the drip irrigation farms from conventional irrigation farms. Hence, the threshold net return per farm due to drip irrigation =₹ 15,292+₹ 9911=₹ 25,203 for drip irrigation farms. The MP of the drip method of irrigation =₹ 1960 per farm of drip irrigation. The MP of the groundwater applied through drip irrigation then =₹ 465+₹ 1960=₹ 2425 per acre-inch. The total net return per farm due to use of groundwater through drip irrigation at the average level of use of groundwater =₹ 15,292+465×28.9 ac-in. per farm+9911×(1)+1960×(1)×(28.9)=₹ 95,285.

Thus, the net return for drip irrigation farms is given by Y=(15,292+9911)+(465+1960)×X inches per farm which yields the function (2) as:

$$Y = 25203 + 2425\mathrm{X}\ \text{for DIF.} \tag{9.2}$$

Table 9.6 Investment behaviour by drip irrigation farms (Tobit model). Dependent variable: investment in rupees per farm on drip irrigation farms

Variable	Coefficient	Standard error	t-value
Intercept	10262[a]	5.12	1967
Net return per farm (₹)	0.23[b]	0.10	2.22
Water use (acre-inches) per farm	−932.96[a]	247	−3.77
Number of observations	68		
Log likelihood function	−401		

[a, b] indicate the significance level at 1 and 5 %, respectively

The negative coefficient for groundwater use per farm (being–₹ 933) indicates that, for every acre inch of groundwater saved on the farm, because of drip irrigation, the investment on drip irrigation increases by ₹ 933

9.3 Net Return Function for Conventional Irrigation Farms

The regression (3) is for farmers using conventional irrigation whose threshold net return=₹ 15,292 per farm reflecting the contribution of all the factors of production other than irrigation water. The MP of groundwater is ₹ 465 per acre-inch at any level of use of water. The total net return per farm due to use of groundwater through flow irrigation at the average level of use of groundwater =₹ 15,292 + 465 × 28.9 ac-in. per farm =₹ 28,730. For 1 ac-in. increase in water use from the mean level, the net returns per farm increases by ₹ 465. The resulting net return for conventional irrigation farms is given by:

$$Y = 15292 + 465\,X \text{ for conventional or flow irrigation farms.} \tag{9.3}$$

9.4 Investment Behaviour of Drip Irrigation Farmers (Tobit)

The investment in drip irrigation made by farmers is regressed on independent variables such as net return per farm (rupees) and water used in acre-inches per farm. The investment on drip irrigation is the actual cost of drip irrigation in drip farms, while it is zero rupees for farms with conventional irrigation. The willingness to pay for drip irrigation is thus estimated using the Tobit maximum likelihood model, where at least one value for dependent variables should be zero (Table 9.6). The results (of the SAS output) indicated that the variables net return per farm (rupees) and water used in acre-inches were significant at 5 and 1 %, respectively. The log likelihood function was significant with a high value of −401. *For every acre-inch of water saved in drip irrigation, the willingness to invest on drip irrigation increases by ₹ 932.*[1] The minimum investment for drip irrigation

[1] The author is grateful to Professor RS Deshpande, Director, Institute for Social and Economic Change, Bangalore, for this interpretation.

Table 9.7 Factors discriminating drip and conventional irrigation farms in the eastern dry zone of Karnataka

Sl. no.	Discriminating variable	Discriminating coefficient (L_i)	Group mean value		$L_i(d_1-d_2)$	Contribution (%)
			DIF members (d_1)	CIF members (control; d_2)		
1	Cropping intensity	0.803	268	265	2.409	0.37
2	Water used in acre-inch	0.283	23	43	5.666	0.88
3	Net returns per acre-inch of water	0.142	5462	964	638.716	98.75

$D^2 = 646.79$

is ₹ 10,262 per farm. The average drip investment per farm was ₹ 41,115 and per acre was ₹ 15,450. For every one rupee increase in net returns per farm, the willingness to pay for drip irrigation increases by 0.23 ₹. *The results amply demonstrate the scarcity value of groundwater as reflected in motivating farmers to invest ₹ 933 on drip irrigation for every one acre-inch of groundwater saved in the process of adoption of drip irrigation.*

9.5 Results of the Discriminant Function Analysis

In order to identify the key variables that discriminate between the DIF and CIF, the stepwise discriminant function analysis is used. The structure matrix indicated that out of the 6 variables considered in the analysis, the variables such as cropping intensity (X_1), water used in acre-inches (X_2) and net returns per acre-inch of water (X_3) were found to be important based on their power to discriminate between DIF and CIF and hence are the significant discriminators between farmers who adopt drip irrigation and conventional irrigation. The estimated function is $Z=0.80\times1+0.28\times2+0.14\times3$. The calculated value of Mahalanobis D^2 is 646.79. In order to find the relative importance of each of the variables in their power to discriminate between the two groups, the percentage contribution of each variable to the total distance measured is worked out (Table 9.7). Thus, the major variable among the three discriminating variables is the *net returns per acre-inch of water* which accounts for 98.75 % of the total distance between the groups. Thus, the farmers shift to drip irrigation, largely considering the net returns they can realize per acre-inch of groundwater. This is an apparent indicator of the farmers' response to the rising cost of groundwater resource due to negative externalities fraught with groundwater irrigation in the hard rock areas in the eastern dry zone of Karnataka.

9.6 Conclusions

There are no compelling reasons to disbelieve that recurrence of drought or excessive rainfall is a phenomenon due to climate change. The manifestation of drought on groundwater wells is both direct and indirect; the direct effect is on crop productivity and increased evapotranspiration. The indirect effects are on the volume of groundwater extracted and declining yield of groundwater from wells over time till such wells get recharged. It is possible that if there is no recharge such wells may be completely dried up. Farmers in contiguous blocks such as Yelluvali and Doddamaralli have adopted drip irrigation en masse; this is a prima facie indicator of the farmers' response to rising costs of groundwater due to negative externalities from cumulative interference of irrigation wells and an attempt to cope with drought (especially since 2001). Even though subsidy for drip irrigation as a governmental programme was extended, since the1980s, farmers selectively began adopting drip irrigation depending upon the crop cultivated. Drip irrigation was first adapted to perennial broad espacement crops such as grapes. Later, drip irrigation was adapted to mulberry. In the years after 2000, farmers in the eastern dry zone began adapting drip irrigation to seasonal commercial crops such as tomato and in floriculture. There are no compelling reasons to believe that large scale adaption of DI is due to subsidy, as the subsidy programme began much earlier. The adaption of drip irrigation is largely due to the effect of economic scarcity of groundwater due to cumulative interference of irrigation wells resulting in high probabilities of initial and premature well failure.

This study has apparently shown that the farmers of the eastern dry agroclimatic zone (especially from Kolar and Chikkaballapura) have demonstrated to the world that drip irrigation is adapted to cultivate even narrow spaced crops due to the rising cost of groundwater and due to negative externalities resulting from cumulative interference among irrigation wells. The farmers thus resorted to drip irrigation due to enhanced MP of water, savings in water use and the net returns per unit volume of groundwater. Despite these benefits and the policy support through subsidy for a DIS, the diffusion of drip irrigation technology for narrow spaced crops in other parts of India is not appreciable. Hence, it is necessary for the departments of agriculture and horticulture in different states to chalk out an intensive outreach programme for expanding drip irrigation for narrow- and broad-spaced crops. In addition, the governments need to keep a close watch on the quality of drip irrigation equipment used by different vendors to save the farmers from poor quality equipment including conveyance tubes and drippers. This calls for developing a cadre of agricultural engineering diploma holders who can be a good human resource to be used by the developmental departments for the benefit of farmers striving hard to realize the economic value of groundwater.

References

Chandrakanth MG, Arun V (1997) Externalities in groundwater irrigation in hard rock areas. Indian J Agric Econ 52(4):761–771

Chandrakanth MG, Romm J (1990) Groundwater depletion in India—institutional management regimes. Nat Resour J 30:485–502

Chandrakanth MG, Priyanka CN, Mamatha P, Patil KK (2013) Economic benefits from micro irrigation for dry land crops in Karnataka. Indian J Agri Econ 68(3):326–38

Nagaraj N, Chandrakanth MG, Gurumurthy (1994) Borewell failure in drought-prone areas of southern India: a case study. Indian J Agric Econ 49(1):101–106

Priyanka CN (2009) Externalities in groundwater use on drip and conventional irrigation farms in the eastern dry zone of Karnataka, MSc(Agri) thesis, unpublished, submitted to the Department of Agricultural Economics, University of Agricultural Sciences, Bangalore

Shashidhara KK, Bheemappa A, Hirevenkanagoudar LV, Shashidhar KC (2007) benefits and constraints in adoption of drip irrigation among the plantation crop growers. Karnataka J Agric Sci 20(1):82–84

Chapter 10
Supply-Side Economic Contribution of Watershed Programme to Groundwater Recharge

10.1 Preamble

Water harvesting for groundwater recharge has been a major objective of watershed programmes in India. The Sujala project initiated by the Government of Karnataka, India, with the assistance of the World Bank is a unique programme, where the project is implemented on both common lands and farmers' lands with cost sharing. Sujala is being implemented in five districts of Karnataka covering 5.11 lakh hectares of land spread over in 77 sub-watersheds, 741 micro watersheds and 1270 villages, benefiting 0.4 million farmers including the ones who are landless. The overall Sujala watershed project cost is ₹ 6777 million, of which ₹ 5408 million is financed by the World Bank, ₹ 725 million is borne by the government and ₹ 643 million is contributed by the farmers. This study aims to assess the economic impact of the Sujala watershed programme and a non-Sujala (Drought-Prone Area Programme—DPAP) watershed in Karnataka on groundwater recharge, agricultural productivity and equity in distribution of benefits among different classes of farmers (Figs. 10.1, 10.2, 10.3).

10.2 Methodology

The main feature of this study is in its estimation of the economic contribution of the watershed programme by comparing the performance in a drought year (2004) with a normal-rainfall year (2005), along with the comparison of the Sujala watershed (with relatively strong institutional background) with a non-Sujala watershed (here, the DPAP watershed) as well as a non-watershed area. We have used the analysis of variance (ANOVA) to reflect the differences.

In the *Veda River* sub-watershed of the Sujala watershed in Hosadurga taluk, one micro watershed, Sivanekatte-1, was selected for detailed study. A non-Sujala (DPAP) watershed in Hosdurga taluk was selected for comparison with the Sujala watershed to estimate the differential impact. A sample of 30 farmers each from

M.G. Chandrakanth, *Water Resource Economics*, DOI 10.1007/978-81-322-2479-2_10

Fig. 10.1 Boulder checks in coconut orchard, Sujala watershed, Karnataka

Fig. 10.2 Spillway constructed for runoff in the Sujala watershed, Karnataka

the Sujala watershed and the non-Sujala watershed and a non-watershed area was chosen, totalling 90 farmers for this study (Seema 2006).

Usually, there are two types of farmer beneficiaries in the watershed programme: (i) farmers totally dependent on rainfall and (ii) farmers possessing irrigation wells

Fig. 10.3 Recharge pit for irrigation borewell in the Sujala watershed programme

as the watershed programme complements the agricultural activities of farmers possessing irrigation wells. It is hypothesized that if the watershed impact is relatively higher on the first category than the second, the purpose of the watershed development programme is served. In this study, the impact of the watershed programme is estimated on these two categories of farmers and in addition estimated for the overall group (i.e. both these classes taken together). In this study, a sample of 30 farmers is drawn from the Sujala watershed programme, another sample of 30 farmers is drawn from a non-Sujala (DPAP) watershed. A third sample of 30 farmers is drawn from a non-watershed-programme area as control for comparison. The required farm-level data have been obtained for the drought year of 2004 and for the normal-rainfall year of 2005 from the Veda River bank sub-watershed and the Shivanekatte micro watershed. The non-Sujala (DPAP) watershed is located in Bokkikere and Srirangapura. The control-area villages are Nagenahally and Honnekere. In each sample, the farmers were post-stratified into two groups of farmers: (i) those who are totally dependent on rainfall and do not possess irrigation wells and (ii) those who possess irrigation wells in the watershed.

10.2.1 Contribution of the Watershed Programme for Farmers Not Possessing Irrigation Wells

Farmers who are totally dependent on rainfall and do not possess irrigation wells form an important class of beneficiaries in a watershed programme. They are far

Table 10.1 Estimated contribution of the Sujala watershed development programme exclusively for farmers who totally depend on rainfed agriculture (and do not possess irrigation wells) on the Veda River bank in Chitradurga district (net returns Rs. per acre)

Sl. No.	Particulars	Drought year (2004)	Normal-rainfall year (2005)
1	Contribution of the non-Sujala (DPAP) watershed programme (= net returns in the non-Sujala WDP minus net returns in the non-watershed area)	4405−4849)=−444	5245−6094=−849
2	Contribution of the watershed institutions (= net returns in Sujala minus net returns in the non-Sujala WDP)	12,203−4405=7798	11,418−5245=6173
3	Contribution of the Sujala watershed (=net returns in Sujala minus net returns in the non-watershed area) =(1)+(2)	12,203−4849)=7354	11,418−6094=5324

DPAP Drought-Prone Area Programme, *WDP* watershed development programme

more exposed to the vagaries of weather and market uncertainties than the 'haves'. The contribution of the Sujala watershed programme for these farmers totally dependent on rainfall is thus a serious equity issue since these farmers with a relatively low endowment will have been benefited the most, compared with the impact on farmers who are in possession of irrigation wells. The contribution of the Sujala and the non-Sujala (DPAP) watershed in a drought year (2004) as well as in a normal-rainfall year (2005) for these farmers is thus estimated using the net returns (as enunciated in Table 10.1).

The estimated contribution of the watershed institutions in the drought year (2004) as well as in the normal-rainfall year (2005) for farmers totally dependent on rainfed agriculture is ₹ 7798 and 6173, respectively. The overall contribution of the Sujala watershed programme to farmers totally dependent on rainfall is ₹ 7354 in the drought year (2004) and ₹ 5324 in the normal-rainfall year (2005). Thus, the Sujala watershed programme has richly benefited the 'have nots' (farmers dependent on rainfall).

In corroboration of these findings, the ANOVA performed by comparing the net returns per acre for farmers dependent on rainfall in a drought year (2004) as well as in normal-rainfall year (2005) in the Sujala watershed, non-Sujala watershed and non-watershed area indicates that the net returns per acre from all sources for farmers totally dependent on rainfall in the Sujala watershed are significantly higher than those in the non-Sujala (DPAP) watershed and in the non-watershed area. Thus, the contribution of the Sujala watershed to farmers totally dependent on rainfall is both statistically and economically significant (Tables 10.4 and 10.5).

Table 10.2 Estimated contribution of the Sujala watershed development programme exclusively for farmers who possess irrigation wells on the Veda River bank in Chitradurga district (₹ per acre)

Sl. No.	Particulars	Drought year (2004)	Normal-rainfall year (2005)
1	Contribution of the non-Sujala (DPAP) watershed programme (= net returns in the non-Sujala WDP minus net returns in the non-watershed area)	6615−5935=680	10,787−5370=5417
2	Contribution of the watershed institutions (= net returns in Sujala minus net returns in the non-Sujala WDP)	6549–6615=−66	10,426−10,787=−361
3	Contribution of the Sujala watershed (= net returns in Sujala minus net returns in the non-watershed area) =(1)+(2)	6549−5935=614	10,426−5370=5056

DPAP Drought-Prone Area Programme, *WDP* watershed development programme

10.2.2 Contribution of the Watershed Programme for Farmers Possessing Irrigation Wells

Considering the contribution of the watershed programme for farmers possessing irrigation wells, the results (Table 10.2) indicated that the contribution of the non-Sujala watershed (DPAP) for farmers possessing irrigation wells is ₹ 680 in a drought year (2004), while it rose to ₹ 5417 in a normal-rainfall year (2005). However, the role of the Sujala watershed institutions is negative in 2004 and 2005 indicating that the institutions have to have different and better strategies exclusively for farmers possessing irrigation wells. This does not mean that the watershed institutions have not performed well. While the watershed institutions have done their best in augmenting incomes of 'have nots' (i.e. those depending totally on rainfall), their role in augmenting incomes of 'haves' has to improve (Table 10.2). Discerning the contribution of the Sujala watershed programme, it is apparent that the overall contribution of the programme for farmers possessing irrigation wells is ₹ 614 per acre in a drought year (2004) and ₹ 5056 per acre in normal-rainfall year. Thus, the contribution of the Sujala watershed as well as the non-Sujala (DPAP) watershed is uniform for the farmers possessing irrigation wells.

10.2.3 Overall Contribution of the Watershed Programme for Farmers Dependent on Rainfall as well as for Farmers Possessing Irrigation Wells

Considering the overall contribution of the non-Sujala (DPAP) watershed on farmers possessing irrigation wells and those not possessing irrigation wells, it was found

Table 10.3 Estimated contribution of the Sujala watershed development programme on the Veda River bank in Chitradurga district (₹ per acre)

Sl. No.	Particulars	2004	2005
1	Contribution of the non-Sujala (DPAP) watershed programme (= net returns in the non-Sujala WDP minus net returns in the non-watershed area)	5689 − 5309 = 380	8246 − 5779 = 2467
2	Contribution of the watershed institutions (= net returns in Sujala minus net returns in the non-Sujala WDP)	8375 − 5689 = 2686	10,746 − 8246 = 2500
3	Contribution of the Sujala watershed (= net returns in Sujala minus net returns in the non-watershed area) = (1) + (2)	8375 − 5309 = 3066	10,746 − 5779 = 4967

DPAP Drought-Prone Area Programme, *WDP* watershed development programme

to be ₹ 380 per acre in a drought year (2004) and ₹ 2467 per acre in a normal-rainfall year (2005). Considering the contributions of the Sujala watershed institutions in a drought year (₹ 2686 per acre) and in a normal year (₹ 2500 per acre) for farmers, the contribution of the watershed institutions is not only uniform irrespective of the agroclimatic conditions but also higher than the contributions of the non-Sujala (DPAP) watershed programme (Table 10.3).

The contribution of the Sujala watershed programme in a normal-rainfall year (2005) being ₹ 4967 is higher than the contribution of the Sujala watershed programme in a drought year (2004) being ₹ 3066 per acre. Thus, the contributions of the Sujala watershed programme in both normal and drought years are higher than the contributions of the non-Sujala (DPAP) watershed as well as the contributions of the Sujala watershed institutions (Table 10.3). Upon performing ANOVA, it is found that the net returns per acre from all sources in the Sujala watershed are significantly different from those in the non-watershed area in a drought year (2004) as well as in a normal year (2005; Table 10.4).

This analysis on incremental net return due to the Sujala watershed pertains to a drought year. With this backdrop, the incremental return in the Sujala watershed has been positive for the sample farmers who do not possess irrigation wells. However, barring the medium farmers, for all sample farmers possessing irrigation wells, the incremental net return per acre is negative. This is because, in the Sujala watershed, the areca nut crop is still in establishment stage. Once the areca nut crop begins bearing, this difference would be positive. When the incremental net return is computed between the Sujala watershed and the non-watershed area, it turns to be positive for sample farmers possessing irrigation wells as well as for those who are totally dependent on rainfall. Here too, the incremental returns are relatively higher for farmers not possessing irrigation wells than for farmers who possess irrigation wells. This reiterates that the Sujala watershed programme has contributed substantially for farmers who are totally dependent on rainfall compared with those farmers who are dependent on irrigation wells (Table 10.5).

Table 10.4 Comparison of the net returns per acre of the Sujala watershed over the non-Sujala (DPAP) watershed and non-watershed area across different categories of farmers at the Veda River, Chitradurga district

Sl. No.	Group(1)	Group(2)	Mean difference (1–2)	Standard error	Sig.	90 % Confidence interval	
						Lower bound	Upper bound
1	*Net returns per acre from all sources for all the farmers in 2004*						
	Sujala watershed	Non-Sujala (DPAP) watershed	12,925.0	6278.5	0.105	−132.5	25,982.7
		Non-watershed area	15,149.5	6278.5	0.047[a]	2091.9	28,207.1
2	*Net returns per acre from all sources for all the farmers in 2005*						
	Sujala watershed	Non-Sujala (DPAP) watershed	11,308.2	6338.9	0.181	−1875.1	24,491.5
		Non-watershed area	13,343.6	6338.9	0.095[a]	160.2	26,526.9
3	*Net returns per acre from all sources for farmers possessing irrigation wells in 2004*						
	Sujala watershed	Non-Sujala (DPAP) watershed	279.4	1988.5	0.989	−3988.9	4547.7
		Non-watershed area	3352.2	2114.7	0.27	−1186.9	7891.4
4	*Net returns per acre from all sources for farmers possessing irrigation wells in 2005*						
	Sujala watershed	Non-Sujala (DPAP) watershed	445.6037	3785.4	0.992	−7679.6	8570.8
		Non-watershed area	2317.557	4025.6	0.834	−6323.3	10,958.4
5	*Net returns per acre from all sources for farmers not possessing irrigation wells in 2004*						
	Sujala watershed	Non-Sujala (DPAP) watershed	20,394.0	9130.3	0.074[a]	1278.2	39,509.8
		Non-watershed area	22,565.2	8925.8	0.037*	3877.5	41,252.9
6	*Net returns per acre from all sources for farmers not possessing irrigation wells in 2005*						
	Sujala watershed	Non-Sujala (DPAP) watershed	17,625.8	9244.4	0.146	−1728.9	36,980.5
		Non-watershed area	19,770.2	9037.4	0.082[a]	848.9	38,691.4

[a] Significant at 90 %

DPAP Drought-Prone Area Programme

Table 10.5 Incremental net returns due to the Sujala watershed over the non-Sujala watershed area and the non-watershed area on the Veda River bank in Chitradurga district in a drought year (2004)

Type of farm	Sujala WDP over the non-Sujala WDP = ₹ 8375–5689= 2686	Sujala WDP over the non-watershed area WDP = ₹ 8375–5309 = 3066		
	For sample farmers possessing irrigation wells	For sample farmers not possessing irrigation wells	For sample farmers possessing irrigation wells	For sample farmers not possessing irrigation wells
Small and marginal farmers	−3782	5863	3618	7714
Medium farmers	2184	7765	3461	6739
Large farmers	−1672	*NA*	1195	*NA*
Overall	−65	7798	614	7354

NA: There were no large farmers in the sample not possessing irrigation wells
WDP watershed development programme

Note Incremental net return in Sujala over the non-Sujala watershed = net return per acre from all sources in Sujala minus that in the non-Sujala watershed.

Considering, sample farmers possessing irrigation wells, their net return is largely influenced by agriculture and horticulture which contribute to 78 % of the net return. The overall net return is ₹ 10,746 which is higher than the one obtained in a drought year, that is ₹ 8375 in the Sujala watershed and ₹ 5689 in the non-Sujala (DPAP) watershed.

Thus, in both years of drought (2004) and normal rainfall (2005), the farmers totally dependent on rainfall (not possessing irrigation wells) are realizing more than 70 % of their net return from livestock, wage income and income-generating activity. This pattern is not very different in the non-Sujala watershed as well as the non-watershed area, where around 50 % of their net return is obtained from livestock, wage income and income-generating activity. This situation gets reversed for farmers possessing irrigation wells.

In the drought year (2004) as well as in the normal-rainfall year (2005) farmers possessing irrigation wells derive at least 80 % of their net return from agriculture and horticulture irrespective of whether they are located in the Sujala watershed, non-Sujala watershed (DPAP) or non-watershed area. Thus, it is apparent that (1) the Sujala watershed benefits are higher for farmers who are totally rainfed over the non-Sujala (DPAP) watershed as well as the non-watershed area. (2) As the farmers possessing irrigation wells are duly engaged in agriculture and horticulture in this farm, their time is unavailable for earning through wage employment, income-generating activity and livestock. (3) Coconut crop occupies at least 70 % of the gross irrigated area for farmers with irrigation wells in the Sujala watershed, non-Sujala watershed and non-watershed area. There were no other major commercial crops cultivated by them other than areca nut (still in bearing stage), onion and groundnut in the Sujala watershed, onion and groundnut in the non-Sujala watershed and onion and sunflower in the non-watershed area which occupy around 10 %.

10.3 Overall Contribution of Watershed and Rainfall

The contribution of the non-Sujala watershed per acre is ₹ 380 in a drought year (2004) and ₹ 2467 in a year of good rainfall (2005). The contribution of the watershed institutions is ₹ 2686 in a drought year (2004) and ₹ 2500 in a good rainfall year (2005). The contribution of the Sujala watershed is ₹ 3066. It is needless to mention that rainfall plays a vital role in shaping the contribution of the watershed programme. The contribution of rainfall to the non-Sujala (DPAP) watershed is ₹ 2087 per acre, and the contribution of rainfall to the watershed institutions is

Table 10.6 Estimated contribution due to the Sujala watershed development programme, institutional innovations, people's participation and rainfall in Chitradurga district (₹ per acre)

Sl. No.	Particulars	Drought year 2004	Normal year 2005
1	Contribution of the non-Sujala (DPAP) watershed programme (= net returns in the non-Sujala WDP minus net returns in the non-watershed area)	5689 − 5309 = 380	8246 − 5779 = 2467
2	Contribution of the watershed institutions (= net returns in Sujala minus net returns in the non-Sujala WDP)	8375 − 5689 = 2686	10,746 − 8246 = 2500
3	Contribution of Sujala watershed (= net returns in Sujala minus net returns in non-watershed area) = (1) + (2)	8375 − 5309 = 3066	10,746 − 5779 = 4967
4	Contribution of rainfall to the non-Sujala (DPAP) watershed programme = (contribution of the non-Sujala (DPAP) watershed in the normal year 2005 minus contribution of the non-Sujala (DPAP) watershed in the drought year 2004)	2467 − 380 = 2087	
5	Contribution of rainfall to the watershed institutions (= contribution of the watershed institutions in the normal year 2005 minus contribution of the watershed institutions in the drought year 2004)	2500–2686 = −186	
6	Contribution of rainfall to the Sujala watershed (= contribution of the Sujala watershed in the normal year 2005 minus contribution of the Sujala watershed in the drought year 2004 is also equal to (4) + (5)	4967 − 3066 = 1901	
7	Net contribution of the non-Sujala (DPAP) watershed	2467 − 2087 = 380	
8	Net contribution of the Sujala watershed	₹ 4967–1901 = 3066	

Note: 2004 was a drought year. 2005 was a good rainfall year. Net effect is the effect of good rainfall
DPAP Drought-Prone Area Programme, *WDP* watershed development programme

₹ −186, that is in a drought year the institutions had to put greater effort. The contribution of rainfall to the Sujala watershed is ₹ 1901 accounting for 38% of the total contribution. The net contribution of the Sujala watershed is ₹ 3066 per acre, while the net contribution of the non-Sujala (DPAP) watershed is ₹ 380 (Table 10.6).

10.4 Implications of the Study

The contribution of the Sujala watershed programme per acre (in normal year, in drought year) for farmers who are totally dependent on rainfall and do not possess irrigation well (₹ 5324 per acre, 7354 per acre) is higher than the contribution of the non-Sujala (DPAP) watershed (₹ −849 per acre, −444 per acre). The contribution of the Sujala watershed programme per acre (in normal year, in drought year) for farmers who are possessing irrigation wells (₹ 5056 per acre, 614 per acre) is almost on par with the contribution of the non-Sujala (DPAP) watershed programme (₹ 5417 per acre, 680 per acre). The contribution of the Sujala watershed programme per acre (in normal year, in drought year) considering all types of farmers (₹ 4967 per acre, 3066 per acre) is higher than the contribution of the non-Sujala (DPAP) watershed programme (₹ 2467 per acre, 380 per acre). The expenditure on the Sujala watershed is around ₹ 4048 per acre, and that on the non-Sujala (DPAP) watershed is ₹ 2632 per acre. If this is assumed to last for 5 years and at a sustainable (or zero) interest rate, the expenditure on the Sujala watershed amounts to ₹ 810 per year per acre, while that on the non-Sujala (DPAP) amounts to ₹ 526 per year per acre. Considering the contribution per acre as benefit and the expenditure per acre as cost, the benefit–cost ratio of the Sujala watershed is 3.78, and that of the non-Sujala (DPAP) watershed is 0.72 in a drought year, while it was 6.13 and 4.69 respectively in a normal year. Thus, the Sujala watershed performed well in both normal and drought years, while the non-Sujala (DPAP) watershed performed well in the normal year.

Reference

Seema HM (2006) Economic impact assessment of Sujala watershed and non-sujala watershed programmes in Chitradurga district. Unpublished MSc(Agri) thesis submitted to the Department of Agricultural Economics, University of Agricultural Sciences, Bangalore

Chapter 11
Water Markets for Sharing Limited Water

11.1 Groundwater Markets (GWMs) for Sharing Limited Water—A Recent Development

About 90% of domestic water needs are met by groundwater in India. In recent years, evidence suggests that groundwater markets (GWMs) are slowly developing as niche water sharing markets endemic in a few pockets. GWMs for non-agricultural purposes, mainly for domestic use, are developing in the semiurban fringes meeting the needs of urban and semiurban areas. The GWM for agricultural purpose is highly localized and is largely need based. The groundwater sold for agricultural purpose is of much larger volume, but the seller realizes much lower value for groundwater than that sold for nonagricultural purpose.

GWMs sharing irrigation water for agriculture generate benefits to buyers, such as higher and risk-free water-income flows thus increasing cropping intensity, changing crop pattern in favour of high value—low water intensive crops and the use of modern farming technologies serving marginal and small farmers with equity. Water markets offer an opportunity to small and marginal farmers who own an irrigation well to increase water utilization for what their own land would not permit, and thereby to spread its overhead cost over a larger command area. The cumulative benefits of water markets are larger than the said effects, the benefits are larger and more widely distributed if water markets are efficient.

Existing literature on GWM is partial as it largely covers the situation in Gujarat (Shah 1991). In addition, most of the work is on the working and structure of GWMs. Some studies have examined the impact of different forms of GWM on crop pattern, productivity, equity and employment. A few studied the reliability and efficiency of GWMs. But, none have quantified the impact of water management under different forms of GWMs.

M.G. Chandrakanth, *Water Resource Economics*, DOI 10.1007/978-81-322-2479-2_11

11.2 GWMs and Efficiency

This chapter explores the efficiency and equity in GWMs by considering sample GWM farmers in hard rock areas (HRAs) of peninsular India (Karnataka) who are sharing groundwater by selling for agricultural purpose, those selling for nonagricultural purpose (mainly for silk filatures/domestic purpose) and comparing their efficiency of water use with farmers who are using their groundwater on their own farm (not involved in sharing/selling groundwater) and farmers who are buying groundwater for agricultural purpose. The Nash equilibrium describing the bargaining power of buyers and sellers of groundwater is also analyzed.

11.3 Hypotheses

Ambiguous property rights and price variation due to end-use pricing characterize natural resource markets. GWMs are no exception to this. This study is aimed to reflect the differential groundwater use efficiencies among farmers with the hypothesis that farmers who use groundwater on their farms and also sell groundwater for nonagricultural purpose are relatively more efficient than (i) farmers who are using groundwater for their agriculture and also sell groundwater for agricultural purpose, (ii) farmers who are buying groundwater for agricultural purpose, (iii) farmers using groundwater for their agriculture and not involved in any selling of groundwater for any purpose. The hypothesis underlying the study is that GWMs bring efficiency in groundwater use in groundwater scarcity areas, since the scarcity rent for groundwater renders farmers to be relatively more efficient than those who are not involved in GWM.

11.4 Sampling

Field impressions pertaining to GWMs activity in Eastern dry zone (EDZ) of Karnataka suggested that the GWM activities are intensive here due to high initial investments for bore wells, high risk of well failure and increasing demand for groundwater for agricultural and nonagricultural purposes. Sidlaghatta taluk where GWM are discernable as compared to other taluks of Kolar district was selected. The taluk was classified as 'dark' area (Department of Mines and Geology, Government of Karnataka, 2001) as more than 85 % of the recoverable recharge was being utilized. The Sidlaghatta Kasaba Hobli was selected for this study as it had concentration of water markets reflected in preliminary survey. Based on the intensity and functioning of water markets, eight villages were selected randomly from the sample hobli. A sample of 120 respondents were selected for detailed economic investigation, comprising following categories[1] (Table 11.1):

[1] (Deepak 2004; Deepak et al. 2005).

Table 11.1 Distribution of sample respondents in groundwater markets (GWMs) in Sidlaghatta taluk, Karnataka (2004)

	FNSW	FSWFA	FBWFA
Village name	No. of sample farmers	No. of sample farmers	No. of sample farmers
Anur	02	02	02
Anur Hunasehalli	05	02	02
Bachalli	06	05	04
Bommanalli	03	03	04
Devaramallur	03	03	03
Pindipapanalli	02	04	04
Sonnenalli	05	03	03
Veerapura	04	08	08
Total	30	30	30

FNSW farmers not selling water, *FSWFA* farmers selling water for agricultural purposes, *FBWFA* farmers buying water for agricultural purposes

FNSW: Farmers Not Selling Water This sample of 30 farmers are those who own irrigation wells, use their water for irrigation and are not involved in either selling or buying groundwater for irrigation.

FSWFA: Farmers Selling Water for Agricultural Purposes This sample of 30 farmers own irrigation wells, use part of this water for irrigating crops and also sells part of the groundwater to the neighbouring farmers for irrigating crops. In turn they receive rent for water let out in terms of crop share of one third to one fourth of the value of gross returns realized by the farmer using the purchased water.

FBWFA: Farmers Buying Water for Agricultural Purposes This sample of 30 farmers buy water for agriculture from their neighbour. Thus, FSWFA and FBWFA are pair of sample farmers of which the first sells and the second buys water for irrigation. Thus, FBWFA may not own any wells, may own irrigation well/s, which may have failed and/or may not be sufficiently yielding groundwater for irrigation.

FSWFNA: Farmers Selling Water for Nonagricultural Purposes This sample of 15 farmers own irrigation wells, use part of this water for irrigating crops and sell part of it for nonagricultural purposes like use in processing of silk yarn by filatures, for small hotels or for domestic use.

BWFNA: Buyers of Water for Nonagricultural Purpose This is a sample of 15 buyers buying water from FSWFNA for purposes like processing silk yarn (filature), hotels or for domestic use (Figs. 11.1–11.12).

Fig. 11.1 Over head tank constructed for filling groundwater to the tankers in water market by SC Deepak

Fig. 11.2 Bullock drawn tanker at a water filling station by SC Deepak

Fig. 11.3 Groundwater purchased for domestic purposes from bullock drawn water tankers by SC Deepak

Fig. 11.4 Beetroot crop grown in water market area by SC Deepak

Fig. 11.5 Tomato—a major vegetable crop grown in water market area by SC Deepak

Fig. 11.6 Ragi crop grown in irrigated condition in groundwater market (GWM) area by SC Deepak

Fig. 11.7 Mulberry—a major crop cultivated by water market farmers by SC Deepak

Fig. 11.8 Silkworm rearing by water market farmer in rearing house by SC Deepak

Fig. 11.9 Groundwater seller and buyer (*first* and *second* from *left*) in seller's field by SC Deepak

Fig. 11.10 Measuring the groundwater yield of the bore well in GWM by SC Deepak

Fig. 11.11 Power tiller drawn water tanker used for conveying-selling water to filatures by SC Deepak

Fig. 11.12 Silk filature utilizing groundwater for boiling silk cocoons by SC Deepak

11.5 Yield of Irrigation Well

Bore well was the predominant groundwater extraction structure in study area. All of the 93 functioning wells in the sample, were bore wells, and there were no functioning dug wells. Yield of bore well was estimated by recording the number of seconds to fill a bucket (of known volume) of water. This was extrapolated to obtain the groundwater yield in gallons per hour (GPH). The yield of bore well in litres is then converted to GPH by using conversion of 4.5 L = 1 gallon.

Volume of groundwater sold for agriculture by FSWFA (= volume bought by FBWFA) in acre-inches is obtained as (yield of well of farmer selling water in GPH × (frequency of irrigation per month of the buyer) × (duration of the buyer's crop in months)/22611

Volume of water sold for nonagricultural purpose by FSWFNA is obtained as (no. of tankers filled per day) × (tanker capacity in litres) × (no. of days of water sold in a month) × (no. of months where water was sold in a year) × (4.54 L/gallon)/22611 gallons per acre-inches.

For FSWFA total groundwater extracted includes water used on their own farm plus the water sold for agriculture to buyers. For FSWFNA, total water extracted includes water used on their own farm plus the water sold for nonagricultural purposes.

11.6 Cost of Production

The Government of Karnataka imposed a flat charge of ₹ 300 per HP per year up to 10 HP pump set since April 1997. Current charges are ₹ 525 per HP per year up to 10 HP. Beyond every 10 HP pump sets farmers are required to install electrical meters. There were no explicit costs of irrigation as there were no operational costs of payments towards electricity utilization earlier. Accordingly, as no variable cost is incurred by farmers in lifting groundwater other than repair and maintenance of the irrigation pump set (IP set) and well there would theoretically be no charges towards the cost of irrigation. However, due to frequent failure of irrigation wells, farmers are forced to invest on additional wells due to externality due to cumulative interference.

Thus the annual cost of irrigation = amortized cost of irrigation well + amortized cost of conveyance + amortized cost of pump set and electrical installation + annual cost of repairs and maintenance. The amortization is over the average life of irrigation wells in each category of sample. The labour cost of irrigation was merged with the cost of other cultural operations. The annual cost of irrigation pertains to each irrigation well on the farm and this is added across all the wells on the farm. This total cost of irrigation is then appropriated over individual crops according to the volume of water used in each crop.

Cost of irrigation per acre-inch = (total amortized cost of irrigation)/(total acre-inches of water used in that year). The cost of irrigation is the cost per acre-inch of water multiplied with the number of acre-inches of water applied to the crop.

The proxy for negative externality (per acre-inch) is taken as amortized cost per acre-inch of water used by FNSW minus amortized cost per acre-inch of water used by intended category of farmers. The amortized cost per well is the total amortized cost of all wells divided by total number of wells. Amortized cost per functioning well is the total amortized cost of all wells divided by total number of functioning wells. The proportion of well failure is equal to the number of failed wells divided by Total number of wells. The net return per acre-inch of groundwater used is equal to total net returns divided by total groundwater used (acre-inches).

Cost of cultivation is the summation of cost of human labour, bullock labour, machine hours, seeds, fertilizers, manure, plant protection, bagging, transporting and packing, the apportioned annual cost of irrigation for each crop on the farm. Proportion of well failure, gross irrigated area (GIA), groundwater used per acre, net returns per acre, amortized cost per functioning a well and all wells, irrigation cost per acre of GIA were calculated to analyze economics of irrigation. Negative externality (per well) = amortized cost per functioning wells minus amortized cost per well.

11.7 Economic Profile of Farmers Involved in Groundwater Market

Considering farmers selling groundwater for agriculture, 40 % of them belonged to marginal and small farmers. However, among farmers selling groundwater for nonagricultural purposes, 70 % were small farmers. Among buyers using groundwater for agriculture, 70 % belonged to marginal and small farmers (Table 11.2). Thus, groundwater sale for agriculture is by both large farmers (60 %) and marginal and small farmers (40 %). Considering buyers of groundwater for agriculture, marginal and small farmers are the predominant. Average size of holding of farmers in GWM is below 5 acres, being 2.17 acres for FSWFNA, 3.45 acres for FBWFA and 4.71 acres for FSWFA. The hypothesis that groundwater sales are usually by large farmers is untenable.

FBWFA had smaller total holding size compared to FSWFA. Moreover, FBWFA had 83 % of total land holding as dry land. The FSWFNA had smaller total holding size, but 74 % of that area was under irrigation. Small land holders with higher proportion of dry land could not afford to invest on well drilling due to their modest resource base, forcing dependency on water markets.

Groundwater sellers sold water to all categories of farmers promoting equity in access to groundwater not only for small farmers, but also for such of those large farmers who could not invest on irrigation wells. Similar results were reported by Prasad (1991) Raju and Rao (1991). According to them, GWMs provided wider access to groundwater to those who were unable to invest. FSWFDP had a smaller

Table 11.2 Socioeconomic features of groundwater seller and groundwater buyer farmers in Sidlaghatta taluk, Karnataka

Sl. No.	Particulars	FNSW (n=30)	FSWFA (n=30)	FBWFA (n=30)	FSWFNA (n=15)
1	Age (years)	45.43	42.2	43.93	38.53
2	Family size (number)	7.47	6.6	5.5	4.63
3	Work force available per family for agriculture (number)	4.07	4.0	4.03	3.93
4	*Schooling*				
	Primary (1–7) (Number)	7 (23.34)	4 (13.34)	10 (33.34)	3 (20.00)
	Secondary (7–10) (Number)	13 (43.34)	21 (70.00)	9 (30.00)	4 (26.67)
	College (Number)	4 (13.34)	2 (6.70)	2 (6.67)	8 (53.34)
	Illiterates (Number)	6 (20.00)	3 (10.00)	9 (30.00)	0 (00.00)
	Total	30 (100)	30 (100)	30 (100)	15 (100)
5	Total land holding size (acres)	6.12 (100)	4.71 (100)	3.45 (100)	2.17 (100)
6	Dry land area (acres)	3.00 (49.00)	2.00 (42.46)	2.87 (83.18)	0.42 (19.35)
7	Area irrigated by irrigation well (acres)	2.23 (36.43)	2.03 (43.09)	0.37 (10.72)	1.60 (73.73)
8	Orchards (mango/eucalyptus)[a]	0.89 (14.54)	0.68 (14.43)	0.21 (6.08)	0.15 (6.91)
9	Marginal farmers (below 2.5 acres)	2 (6.67)	4 (13.34)	8 (26.67)	3 (20)
10	Small farmers (2.5–5 acres)	8 (26.67)	8 (26.67)	13 (43.34)	9 (60)
11	Large farmers (above 5 acres)	20 (66.66)	18 (60)	09 (30)	03 (20)
12	Total	*30(100)*	*30(100)*	*30(100)*	*15(100)*

Note: Figures in parentheses show percentage to total

FNSW farmers not selling water, *FSWFA* farmers selling water for agricultural purpose, *FBWFA* farmers buying water for agriculture, *FSWFNA* farmers selling water for nonagricultural purposes

[a] Considering that 12, 10, 3 and 2 farmers in FNSW, FSWFA, FBWFA and FSWFNA had mango/eucalyptus orchard, the area is 2.22, 2.04, 2.10 and 1.13 acres per farm respectively

holding size of 2.17 acres and still they invested in irrigation wells, as they were located in the city outskirts and had other sources of income, better knowledge and were well informed.

11.8 Features of Irrigation Wells

In the case of FNSW, FSWFA, FSWFDP all the farmers owned functioning bore wells but in the case of FBWFA only five sample farmers (17 %) owned functioning bore wells (Table 11.3). The sample of 30 FBWFA had eight irrigation wells of

Table 11.3 Particulars of irrigation wells of sample farmers in GWM in Sidlaghatta taluk, Karnataka (2003–2004)

Particulars	FNSW (n=30)	FSWFA (n=30)	FBWFA (n=30)	FSWFNA (n=30)
Number of farmers owning functioning well	30	30	05	15
Total number of functioning wells	38 (82.60)	35 (83.40)	05 (62.50)	15 (83.40)
Total number of nonfunctioning wells (includes dug wells and borewells)	08 (17.39)	07 (16.67)	03 (37.50)	03 (16.67)
Total number of bore wells	46 (100)	42 (100)	08 (100)	18 (100)
Average age of wells as on 2003 (years)	10.14	9.23	8.78	8.86
Gross irrigated area (GIA) using water from own wells (acres)	191.5	186	24.5	73
GIA per functioning well	5.03	5.31	4.60[a]	4.66
Average number of all wells per farm	1.54 (3 wells for 2 farms)	1.40 (4 wells for 3 farms)	0.26 (1well for 4 farms)	1.20 (5 wells for 4 farms)
Average number of functional wells per farm	1.27 (5 wells for 4 farms)	1.17 (7 wells for 6 farms)	0.17 (1 well for 6 farms)	1.0 (1 well for 1 farm)
Average depth of the well (feet)	375.55	420.30	381.24	362.67
Average yield of wells (gallons/hr)	2500	2900	1550	2750
Number of farms without functioning wells	00	00	25	00
Amortized cost (Rs) per irrigation well	14,058	15,547	13,483	15,427
Amortized cost (Rs) per functioning irrigation well	15,725	17,417	17,544	16,885

Note: Figures in parentheses show percentage to total

FNSW farmers not selling water, *FSWFA* farmers selling water for agricultural purpose, *FBWFA* farmers buying water for agriculture, *FSWFNA* farmers selling water for nonagricultural purposes

[a] Includes area irrigated through purchased water and from own wells

which five (62% of) wells were functioning. Thus, FBWFA suffered the highest failure rate of wells and eventually resorted to buying groundwater from neighbouring farmer-seller. Even though GIA per well is around 5 acres and average age of wells was 9–10 years, for all categories of farmers, wide variation existed in the total area irrigated by all wells under each category as the number of functioning wells varied from 8 wells for 30 FBWFA to 46 wells for 30 FNSW. The negative externality per irrigation well was the highest for FBWFA (₹ 4061), whereas for other categories it ranged from ₹ 1500 to 1900. There were 25 FBWFA whose main source of irrigation water was the purchased water, as these farmers did not own any well. Considering this predicament, the plight of FBWFA has been reduced by FSWFA resulting in equity through GWM.

11.9 Cropping Pattern

Mulberry a hardy low water perennial is the major crop preferred by farmers in GWM. The choice of mulberry crop by farmers in water market in itself is a prima facie indicator of the prevailing groundwater scarcity in the eastern dry zone of Karnataka. The FBWFA had 87 % of the GIA under mulberry cultivation followed by FSWFA (70 %), FSWFDP (64 %) and FNSW (56 %) (Table 11.4). The area under other crops such as tomato, chillies, carrot, beetroot and potato, which require relatively more water, was higher for FNSW over other categories of farmers. While, there was no conditionality of FSWFA on FBWFA regarding crops to be irrigated with purchased water, both buyers and sellers were relatively efficient in using water compared to FNSW. The area irrigated under perennials was higher for FSWFA compared to FBWFA and thus, water sellers have reliable and sustainable sources of income with better resource endowment than buyers.

11.10 Sources of Income for Farmers in Groundwater Market

For the three groups of farmers participating in GWM (FSWFA, FNSW, FSWFNA), income from farming ranges between 20 and 30 % of the annual income, whereas for FBWFA, the crop farming contributes 15 % of their annual income (Table 11.5). While, FSWFA realized 17 % of their annual income from sale of groundwater, FSWFNA realized 41 %. FSWFA not only sold higher volume of 47 acre-inchesper farm for agriculture, but also realized lower returns (of ₹ 29,069 per farm), when compared with FSWFNA who sold half the volume sold by FSWFA (25 acre-inches) but realized an income (₹ 85,000) twice that of FSWFA. This illustrates that end-use price of the resource and not so much the volume of water sold, is the key for shaping marginal productivity of groundwater. Groundwater for silk filatures and domestic use fetched much higher price compared to agriculture. Thus, the sale of groundwater added considerably to the total net returns of the FSWFA. This sale of groundwater which added around 17 % of the gross returns is equivalent to the contribution of dairy income.

Considering the net returns from different sources of income of FBWFA, FNSW and FSWFA, farmers derived major part of the returns from sericulture, which contributed 30–40 % of the total net returns. Livestock rearing is main source of income for FBWFA, which adds 46 % to the total net returns.

Income from dry land, perennial crops like mango were not considered here, as the major focus of the study was dynamics of water markets.

Table 11.4 Cropping pattern of different categories of sample farms in GWM in Sidlaghatta taluk, Karnataka (2003–2004) (area in acres)

Crops (Season Wise)	FNSW (n=30)		FSWFA (n=30)		FBWFA (n=30)		FSWFNA (n=15)	
	Net cropped area	Percent to total	Net cropped area	Percent to total	Net cropped area	Percent to total	Net cropped area	Percent to total
Kharif								
Tomato	14.5	7.57	08.5	4.56	03.0	2.31	04.5	6.16
Chillies	03.0	1.56	07.5	4.03	00.0	00.0	00.0	00.0
Carrot	02.0	1.04	03.5	1.88	00.0	00.0	03.0	4.10
Beet root	10.0	5.22	0.00	00.0	02.0	1.54	00.0	0.00
Maize	5.50	2.87	00.0	00.0	00.0	00.0	00.0	0.00
Ragi	00.0	00.0	00.0	00.0	03.0	2.31	00.0	0.00
Potato	03.0	1.56	00.0	00.0	02.0	1.54	02.5	3.42
Rabi								
Tomato	06.0	3.13	06.0	3.22	02.0	1.54	02.0	2.73
Ragi	00.0	00.0	02.0	1.07	00.0	00.0	00.0	0.00
Potato	14.0	7.31	03.0	1.61	02.0	1.54	03.5	4.79
Carrot	02.0	1.04	05.5	2.95	1.5	1.18	02.5	3.42
Beet root	06.0	3.13	00.0	00.0	02.0	1.54	00.0	0.00
Maize	02.0	1.04	00.0	00.0	00.0	00.0	00.0	0.00
Summer								
Tomato	12.0	6.26	14.5	7.79	01.5	1.18	04.5	6.16
Carrot	02.0	1.04	00.0	00.0	01.5	1.18	02.0	2.73
Ragi	02.0	1.04	00.0	00.0	02.0	1.54	01.5	2.05
Chillies	05.0	2.61	03.0	1.61	02.0	1.54	00.0	0.00
Annuals								
Mulberry[a]	35.5	55.61	43.5	70.16	37.5	86.87	15.5	63.69
GIA	191.5	100	186	100	129.5	100	73.0	100
Perennials								
Mango[a]	20.0	75.47	14.5	70.73	02	30.76	00.0	00.0
Eucalyptus[a]	6.50	24.52	06	29.26	4.5	69.23	03.0	100
Total perennial	26.5	100	20.5	100	6.5	100	03.0	100
Rain fed								
Ragi	58.0	62.36	35	59.32	49	56.98	04.0	61.53
Maize	35.0	37.63	24	40.67	37	43.07	02.5	38.46
Rain fed total	93.0	100	59	100	86	100	6.5	100

FNSW farmers not selling water, *FSWFA* farmers selling water for agricultural purpose, *FBWFA* farmers buying water for agriculture, *FSWFNA* farmers selling water for nonagricultural purposes

[a] For Mulberry, 1 acre of crop is considered as 3 acres, for mango and eucalyptus, 1 acre of crop is considered as 2 acres while computing cropping intensity

Table 11.5 Sources of farm income in different categories of sample farms in GWM in Sidlaghatta taluk, Karnataka (2003–2004) (net returns in ₹ per farm per year)

Source of income per farm	FNSW (n=30)	FSWFA (n=30)	FBWFA (n=30)	FSWFNA (n=15)
From farming (irrigated agriculture)	47,149 (30.98)	41,584 (24.58)	10,683 (14.90)	43,783 (20.96)
From dairy	26,884 (17.67)	29,485 (17.42)	32,965 (46.01)	9154 (4.38)
From sericulture	61,426 (40.37)	58,691 (34.69)	20,431 (28.51)	45,500 (21.78)
From sale of groundwater	–	29,069 (17.18) from selling 1396 acre-inches for irrigation	–	85,000 (40.69) from selling 368.46 acre-inches for nonagricultural purpose
From other sources (business, job)	16,666 (10.95)	10,334 (6.10)	7,563 (10.56)	25,428 (12.17)
Total annual income per farm	152,125 (100)	169,163 (100)	71,642 (100)	208,865 (100)

Note: Figures in parentheses show percentage to total.
FNSW farmers not selling water, *FSWFA* farmers selling water for agricultural purpose, *FBWFA* farmers buying water for agriculture, *FSWFNA* farmers selling water for nonagricultural purposes.

11.11 Economics of Irrigation

The net return per acre of GIA is the highest being ₹ 18,345 for FSWFNA followed by ₹ 17,009 for FNSW, ₹ 16,173 for FSWFA and ₹ 7208 for FBWFA (Table 11.6). FBWFA are relatively more efficient in using water for irrigation than FSWFA, as they used 15.41 acre-inches when compared to 21.21 acre-inches used by FSWFA per acre of GIA. Both FSWFNA and FBWFA were the most efficient in using groundwater for irrigation, as they were judicious in using scarce water in irrigating crops at their critical stages of growth giving protective irrigation using around 15 acre-inches per acre. This was due to their awareness of the marginal productivity of groundwater due to their exposure to GWM. As mentioned earlier, end-use pricing prevails in price differentiation for natural resources and groundwater is no exception to this. In corroboration, the net return per acre-inch of groundwater used for irrigation was the highest for FSWFNA, followed by FNSW, FSWFA and FBWFA in that order.

11.12 Income from Sale of Groundwater

Considering the options for the groundwater farmer to use as well as to sell groundwater, FSWFA realized ₹ 762 per acre-inch of groundwater, while they realized ₹ 624 for every acre-inch of groundwater sold, hardly a difference of ₹ 138 per acre-inch (Table 11.7).

Table 11.6 Economics of irrigation in different categories of sample farms in GWM in Sidlaghatta taluk, Karnataka (2003–2004)

Sl. No.	Particulars	FNSW ($n=30$)	FBWFA ($n=30$)	FBWFA ($n=30$)	FSWFNA ($n=15$)
1	Number of sample farmers	30	30	30	15
2	Cropping intensity (percent)	174.8	201.7	163	244.2
3	Irrigation intensity (percent)	260	286	272	287
4	GIA (acres)	191.5	186	129.5	73
5	GIA per farm (acres)	6.38	6.20	3.65	5.54
6	GW extracted and used for farming from own wells (acre-inches)	4949	3946	592.25	1100
7	GW Marketing (acre-inches)	–	1395.75 sold	1395.75 purchased	368.46 sold
8	GW Marketing per acre of GIA (acre-inches) (=7/4)	–	7.50 sold	10.77 purchased	5.04 sold
9	Total GW extracted/used including purchased/sold groundwater (acre-inches) (=6+7)	4949	5341.75	592.25+1395.75=1988	1100+368.46=1468.46
10	Water extracted per well (acre-inches)	107.58	127.16	74.00	81.55
11	GW used on per farm basis (acre-inches)	164.96	131.53	66.26	36.66
12	GW used per acre of GIA (acre-inches) (=9/4)	25.91	21.21	15.41	15.64
13	Amortized cost of irrigation per acre-inch of groundwater from owned irrigation well (₹)	131	122	182	189
14	Amortized cost per well (₹)	14,058	15,547	13,483	1542
15	Total net returns of sample farms (₹)	3,257,250	3,008,250	933,420	1,339,245
16	Net returns per farm (₹)	108,575	100,275	31,114	89,283
17	Net returns per acre-inch of Groundwater used (₹/acre-inch) (=15/9)	658	762	470	1217
18	Net returns per acre of GIA (₹)	17,009	16,173	7208	18,345

FNSW farmers not selling water, *FSWFA* farmers selling water for agricultural purpose, *FBWFA* farmers buying water for agriculture, *FSWFNA* farmers selling water for nonagricultural purposes

Table 11.7 Net returns from agriculture and from sale of groundwater in GWM in Sidlaghatta taluk, Karnataka (2003–2004)

Particulars	FSWFA (*n*=30)	FBWFA (*n*=30)	FSWFNA (*n*=15)
Total net returns on sample farms (₹)	3,008,250	933,420	1,339,245
Net returns per farm (₹)	100,275	31,114	89,283
Total net returns including returns from sale of water among sample farms (₹)	3,880,320	–	2,614,245
Net returns per farm including returns from sale of groundwater (₹)	129,344	–	174,283
Net returns (per farm) from selling groundwater per year (₹)	29,069	–	85,000
Total groundwater extracted including sold/bought water on all sample farms (acre-inches)	5341.75	1988	1468.46
Total groundwater extracted and used for farming from own irrigation wells on all sample farms (acre-inches)	3946	592.25	1100
Total groundwater purchased/sold (acre-inches) on all sample farms	1395.75	1395.75	368.46
Net returns per acre-inch of water from farming (₹/acre-inch)	563	470	912
Net returns per acre-inch of groundwater from farming and sale of groundwater (₹/acre-inch)	726	–	1780
The addition to the net returns per acre-inch of groundwater extracted (₹)	163	–	868
Net returns per acre-inch from using own water in farming (₹/acre-inch)	762	–	1217
Notional price realized per acre-inch of groundwater sold (₹/acre-inch)	624[a]	–	6910[b]

FNSW farmers not selling water, *FSWFA* farmers selling water for agricultural purpose, *FBWFA* farmers buying water for agriculture, *FSWFNA* farmers selling water for nonagricultural purposes

[a] Table (4.15) offers greater details

[b] Table (4.16) offers greater details

This shows that FSWFA, by selling their groundwater, realized almost the same amount of return (₹ 624) as they could realize by using groundwater on their farm (₹ 762). Thus, FSWFA gets an additional ₹ 624 per acre-inch purely by selling groundwater and this additional ₹ 138 per acre-inch is not large enough to cover the additional risks and uncertainty in farming. FSWFNA are also realizing higher net return from selling. The notional price realized by FSWFA and FSWFNA per acre-inch of groundwater is ₹ 624 and 6910 respectively. This whopping difference of ₹ 6286 is due to end-use pricing.

Considering the net return of ₹ 7208 per acre accounting for the cost of purchased water, the cost of purchased water for agriculture forms about 50 % of the net returns (Table 11.8). It is in order to appreciate the contribution of purchased groundwater for FBWFA, who are irrigating 4.29 acres using purchased water for every acre of irrigated area from their irrigation well. On an average, a FBWFA

Table 11.8 Economy of groundwater buyers for irrigation in Sidlaghatta taluk, Karnataka (2003–2004)

Sl. No.	Particulars	Quantity
1	Number of farmers buying groundwater for irrigation *(n)*	30
2	GIA of all sample buyers (acres)	129.50
	(A) Gross area irrigated with purchased groundwater (acres)	105 (81)
	(B) Gross area irrigated with groundwater from own irrigation wells (acres) (only five farmers among FBWFA owned irrigation wells)	24.50 (19)
	(C) Area irrigated with purchased groundwater per acre of area irrigated using own groundwater	4.29
3	Average GIA irrigated from purchased water (acres)	3.5
4	Net returns per acre-inch of groundwater used (₹/acre-inch)	470
5	Payment made for groundwater purchased per farm per year (₹)	34,745
6	Net returns per farm per year from agriculture after paying for purchased groundwater (₹)	31,114
7	Net returns per acre of GIA per year from agriculture after paying for purchased groundwater (₹)	7208

Note: Figures in parenthesis indicates percent

realizes a net return of ₹ 7208 per acre of GIA. The irrigation cost is the highest for FBWFA being two to four times higher than the cost incurred by other categories of farmers. When this cost is compared with the risk of initial well failure and the relatively higher investment and the interest payments on borrowed investment, FBWFA felt it most appropriate to buy water for irrigation. Considering the sale of groundwater by FSWFA and FSWFNA, both the categories sold about 25 % of total groundwater extracted by them but realized differential net return due to the price differential obtained, a strong function of the end use. The FSWFA and FSWFNA sold 46.52 and 24.56 acre-inches of groundwater per farm, and 7.5 and 5 acre-inches per acre of their GIA.

11.13 Groundwater Sold for Different Purposes

Considering the two groups of farmers FSWFA and FSWFNA selling groundwater, both groups of farmers sold around 25 % of the water extracted, but realized different prices depending on the end use to which groundwater was sold (Table 11.9). The estimated price realized by FSWFA is ₹ 624 per acre-inch as they sold for irrigation, whereas FSWFNA realized ten times this price, since they sold groundwater for nonagricultural purposes (silk filatures/domestic use). The FSWFNA extracted around 45 % lower groundwater extracted by FSWFA, but realized 192 % higher net return than FSWFA. Thus, end use and end-use price play a key role in determining the volume of groundwater extracted, sold and the revenue realized in water market.

Table 11.9 Details of groundwater sold for different purposes in GWM in Sidlaghatta taluk, Karnataka (2003–2004)

Particulars	FSWFA (n=30)	FSWFNA (n=15)
Total water extracted by all sample farms (acre-inches)	5341.75	1468.46
Groundwater extracted per farm (acre-inches)	178.05	97.89
Groundwater extracted per acre of GIA (acre-inches)	29	20
Total volume of groundwater sold (acre-inches)	1395.75	368.46
Volume of groundwater sold per farm (acre-inches)	46.52	24.56
Volume of groundwater sold per acre of GIA (acre-inches)	7.50	5.04
Proportion of groundwater sold (percent) out of groundwater extracted	26.12	25.09
Net return (per farm) from selling groundwater (₹)	29,069	85,000
Estimated price realized per acre-inch of groundwater (₹ per acre-inch)	624	6910
Price realized per acre-inch of groundwater sold for nonagricultural use per rupee of price per acre-inch of groundwater sold for agricultural purpose	–	11.07

FSWFA farmers selling water for agricultural purpose, *FSWFNA* farmers selling water for nonagricultural purposes.
FSWFNA selling mainly for silk filatures and for domestic purposes, as well as FSWFA sold around 25 % of their groundwater extracted

11.14 Institutional Arrangements in Groundwater Market

The only mode of groundwater sale and method of payment for agriculture in the groundwater niche market was through crop share contract (Saleth 1994). The groundwater price ranged from one third to one fourth of the gross value of produce cultivated using the purchased groundwater. About 90 % of the buyers opted for 1-year contract and 10 % opted for seasonal contract. Easy monitoring of groundwater supplies, payment, recovery that lead to little or no disputes, were the major reasons for preferring crop share method of payment for groundwater. In addition, this method of share crop contractual payment for groundwater is in vogue in other parts of Kolar district where groundwater niche markets have been in existence since the last 10 years as opined by the farmers. About 17 % of the sellers also indicated that crop share method offered them remunerative net return to groundwater. However, they did not indicate other alternative modes of groundwater pricing.

11.15 Estimated Price of Groundwater in Nice Water Markets According to End Use

Considering the estimated price of groundwater in agriculture (Table 11.10), in the crop share contract basis, the groundwater price ranged from ₹ 567 to 656 per acre-inch. At the average amortized cost of ₹ 122 per acre-inch of groundwater,

Table 11.10 Estimated price of groundwater in different crops in GWM, Sidlaghatta taluk, Karnataka (2003–2004)

Crop	Yield of the crop (tons/acre)	Groundwater price as crop share (or proportion of crop yield)	Quantity of crop share (tons/acre)	Groundwater bought and used per acre (acre-inches)	Estimated value of groundwater used per acre as equal to Market value of the crop share quantity (₹)	Estimated price of groundwater sold for agriculture in ₹ per acre-inch	Amortized cost per acre-inch of groundwater	Estimated price—amortized cost ratio
Mulberry	8.56	1/3	2.85	5.4	3420	3420/5.4 =633	122	5.19
Tomato	23.30	1/4	5.82	19	12460	12460/19=656	122	5.37
Potato	6.21	1/4	1.55	12	6800	6800/12 =567	122	4.65
Carrot	6.06	1/4	1.51	11	6795	6795/11 =617	122	5.06
Beetroot	6.82	1/4	1.71	9.5	6014	6014/9.5 =633	122	5.19

the estimated price to amortized cost ratio ranged from 4.65 to 5.37. The price of groundwater for nonagricultural purposes, ranged from ₹ 4 to 6.9 per 100 L of water for purposes like household use, hotels and silk filatures (Table 11.11).

11.16 Nash Equilibrium Model of Groundwater Pricing

The absence of competitive markets in transaction of groundwater between owners of private irrigation systems and nonowners is widely observed. Topography constrains the deliverable area of water. Capacity of water pump, length of conveyance facilities, such as pipelines also limits the deliverable area. Thus, the number of potential buyers that the owners of irrigation systems can expect is physically limited, and the number of potential sellers that the nonowners can expect as well. Usually the owner farmer and nonowner farmer in close location is a pair of seller and buyer for water transactions. The price is determined under bilateral bargaining between them (Kei and Takeshi 1999). Thus, the price is not necessarily a competitive one. There is also a concern about water seller's exploitative pricing behaviours towards buyers as the owner is treated as water lord.

Assumptions of Nash equilibrium:

- The two players are highly rational.
- Both have equal bargaining skills.

Table 11.11 Price of groundwater purchased for nonagricultural purposes from farmers in GWM in urban fringe of Sidlaghatta town Karnataka (2003–2004)

Groundwater buyers	Number of groundwater buyers in the sample (n=15)	Groundwater purchased in litres per day	Groundwater bill (₹/day)	Price per litre of groundwater (₹)	Price per 100 L of groundwater (₹)	Justification regarding the groundwater price
Silk filatures	08 (53.66)	1400	95	0.069	6.90	Power tiller driven tankers are the main mode of transporting water with tanker capacity ranging from 1400 to 1500 L per tanker
Hotels	04 (26.34)	1100	70	0.064	6.40	Bullock drawn, cart mounted tankers are the main mode of transport, the tanker capacity ranging from 250 to 375 L
Households	03 (20)	400	20	0.040	4.00	Bullock drawn, cart mounted tankers are the main mode of transport with tanker capacity ranging from 250 to 375 L

Note: Price of the tiller mounted tankers ranges from ₹ 90 to 100 per load.

- Each has full knowledge of the tastes and preferences of the other.
- Each individual wishes to maximize the utility to himself of the ultimate bargain.

According to Nash equilibrium, bargaining power among the buyers and sellers of groundwater is the key determinant in GWM. Accordingly, if the bargaining power of the seller is measured in terms of two variables namely GIA of sellers and total water extracted by sellers, bargaining power of buyer is measured by the GIA of buyers. However, as crop share is one third to one fourth across different crops, the dependent variable of water price tends to be uniform. Thus, for performing regression analysis, variation in groundwater price as induced by price-cost ratio as equal to price per acre-inch of groundwater charged by the sellers divided by the amortized cost per acre-inch of groundwater of the seller, is considered as the dependent variable. This price-cost ratio is taken as a surrogate for market price. Nash bargaining model is thus constructed as:

Ratio of groundwater price to its amortized cost = f (GIA of sellers, total water extracted by sellers, GIA of buyers). Empirically Nash bargaining model was assumed to follow the quadratic price-cost ratio *(Y)* function: $Y = a + b_1x_1 + b_2x_2 + b_3x_3 + c_1x_1^2 + c_2x_2^2 + c_3x_3^2$, where, x_1=GIA of seller; x_2=Total water extracted by seller, x_3=GIA of buyer;

Table 11.12 Nash equilibrium model of groundwater niche market dependent variable: Ratio of groundwater price to its amortized cost ($n=30$)

	Intercept	Inputs						R^2
		GIA of seller (X_1)	Total water extracted by seller (X_2)	GIA of buyer (X_3)	X_1^2	X_2^2	X_3^2	
Estimated coefficient	−1.0938	2.6861 (1.7440)	−0.1334 (0.0884)	4.1523 (2.1150)	−0.208 (0.1064)	0.0004 (0.0002)	−0.4928 (0.245)	0.7328
t value		1.5402	−1.5079	1.963	−1.960	2.313	−2.009	

Note: Figures in parentheses are standard errors

Since price-cost ratio is a dependent variable, elasticity of price-cost ratio with respect to each of the explanatory variables is considered as $(b_i + 2c_i \overline{x_i})\overline{x_i} / \overline{Y_i}$ (Table 11.12).

The results of the above Nash equilibrium model indicated that for 1 % increase in GIA of the seller, the price-cost ratio increased by 0.098 %. For 1 % increase in total water extracted by sellers, price-cost ratio increased by 0.85 % and for 1 % increase in GIA of buyers, the ratio increased by 0.3 %. Thus, total water extracted by sellers and GIA of buyers are the key explanatory variables in price determination in GWM.

The total water extracted by sellers gives greater bargaining power for the seller in price determination than the GIA of buyers and GIA of the seller. This also shows that the groundwater seller by judiciously distributing the extracted water between his farm and neighbouring buyer farm is able to reap larger benefits to the tune of ₹ 29,069 per farm.

11.17 Implications

Considering the volume of groundwater used per acre of GIA, it was considerably lower for farmers involved in water market viz. 21 acre-inches for FSWFA, 15 acre-inches for FBWFA and 16 acre-inches for FSWFNA when compared to 26 acre-inches for FNSW. At the same time, the net returns per acre-inch of groundwater used for irrigation was higher for farmers involved in GWM, viz., ₹ 762 for FSWFA, ₹ 908 for FBWFA and ₹ 1217 for FSWFNA when compared to ₹ 658 for FNSW.

These outcomes indicate that the water markets have helped the farmers in using the groundwater more efficiently. In addition (25) farmer water buyers not owning any irrigation bore wells could buy water for irrigation realized higher net returns, than those owning irrigation wells, and 70 % of these buyer farmers were marginal and small farmers, is an equity implication.

GWMs have the potential to promote further water use efficiency among sellers and buyers by installing drip/sprinkler systems, since the seller is aware of higher marginal productivity of water and the buyer is aware that he has to pay for his water through his nose.

There has been increased extraction of groundwater in this region. This has repercussions on sustainability of groundwater use. Hence, to promote sustainability of groundwater use and thus increasing the accessibility of water to farmers, groundwater literacy should be promoted including: (i) role of desiltation and tank rehabilitation process in augmenting groundwater recharge in HRAs, (ii) type of well to be constructed, the depth and volume of availability of groundwater, in relation to the existing and projected interference problems in aquifers where groundwater mining is apparent, (iii) quality, type and HP of the pump to be used, (iv) cropping pattern to be followed in relation to water requirement of crops and the volume of availability of groundwater every year.

By selling water for agricultural and nonagricultural purpose, farmer realized higher marginal productivity of water when compared with farmers who did not sell water. Even with this realization of higher marginal productivity of water, farmers who sold water for agricultural or nonagricultural purposes realized 17 and 41 % of their total net return from sale of water respectively. This is an additional income not at the sacrifice of their original returns from agriculture. Thus, GWM in groundwater scarce area has not only helped farmers to achieve economic water use efficiency, but also complemented additional net return per unit of water extracted. Therefore, GWMs in rural areas can be facilitated by considering GWMs as legal with well defined property rights to water.

Farmers, who are able to sell their groundwater irrespective of the purpose, are relatively more efficient than farmers who are just using groundwater for irrigating their own crops. It is crucial to note, however that, farmers selling groundwater have not abandoned their agriculture, and have continued farming in addition to groundwater sale. Thus, farmers participating in GWM are more efficient in water use in addition to conserving groundwater than those who are not participating in GWM. Thus, GWMs are promoting efficiency in groundwater use and in addition are a conservation strategy for the scarce groundwater in HRAs.

References

Deepak SC (2004) An economic analysis of groundwater markets in Eastren Dry zone of Karnataka, Un published MSc(Agri Econ) thesis, Dept of Agri Economics, UAS, GKVK, Bangalore 560065

Deepak SC, Chandrakanth MG, Nagaraj N (2005) Groundwater markets and water use efficiency: the case of Karnataka. Water Policy Research Highlight No. 12, IWMI-TATA Water Policy Program, 5th Annual IWMI-TATA partnerships meet, Feb 2005

Kei K, Takeshi S (1999) Price determination under bilateral bargaining with multiple modes of contract: a study of groundwater market in India. (from google.com)

Saleth RM (1994) Water institutions in India–economics, law and policy. Commonwealth publishers, New Delhi

Shah T (1991) Water markets and irrigation development in India. Indian J Agric Econ 46(3): 335–348

Chapter 12
Sustainable Path of Extraction of Groundwater in Tank and Canal Command Areas

Preamble The optimal use of a natural resource under optimal control regimes arises from the fact that the farmer considers the user cost of the resource, unlike in competitive (no control) regime, where the marginal cost is equated to the current marginal benefit in determining the optimal level of resource extraction. Hence, the myopia of ignoring user cost under competitive regime leads to overexploitation causing intergenerational inequity in the groundwater resource. The control variable (decision variable) is the volume of water to be pumped in each period. The state variable is the groundwater resource stock in each time period. The objective function is the net social benefit for the given groundwater to be maximized subject to the stock of groundwater influenced inter alia by cost of externality, pumping lift and electricity cost.

12.1 Model

The primary data for this study are drawn from 30 farmers possessing groundwater wells under sole irrigation (GWSI; without any recharge); 30 farmers having groundwater wells under tank irrigation (GWTI) command and 30 farmers having groundwater wells under canal irrigation (GWCI) command (Figs. 12.1, 12.2, 12.3, 12.4, 12.5) in Belgaum district, Karnataka.[1] In the model, the volume of groundwater to be pumped in each period is the control variable, while the state variable is the groundwater stock in each period. λ is the co-state variable measuring the user cost or groundwater rent reflecting the net benefit (NB) foregone in future due to the present use of groundwater. $P=a-2bw$ indicates the inverse groundwater demand function, where P is the groundwater price and w is the volume of groundwater pumped (acre-inches or ha cm); a and b are parameters of the inverse demand function.

[1] The results of this chapter are drawn from Rajendra (2003).

M.G. Chandrakanth, *Water Resource Economics*, DOI 10.1007/978-81-322-2479-2_12

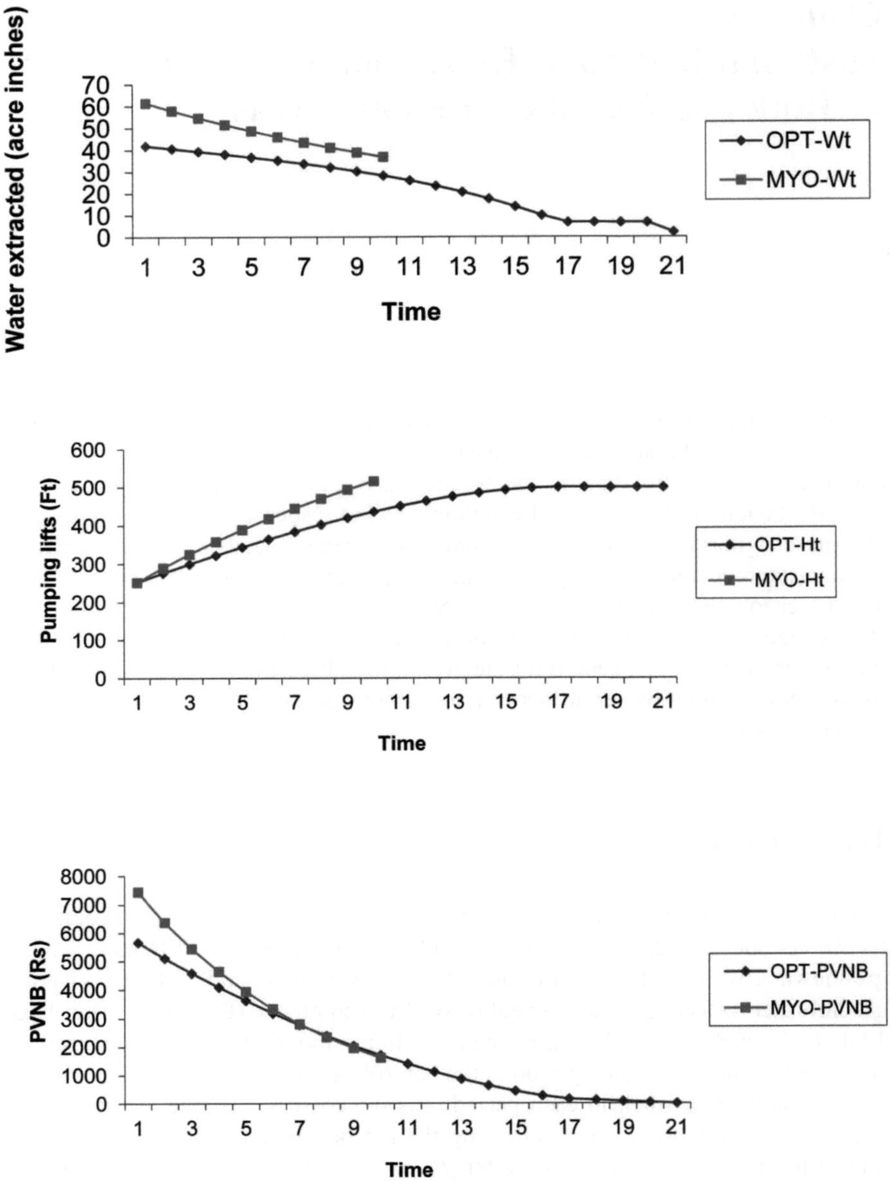

Fig. 12.1 Farms cultivating **a** sunflower and **b** tomatoes under sole well situation and **c** paddy under tank command in Ramadurga taluk, Belgaum district. (Source: Rajendra 2003)

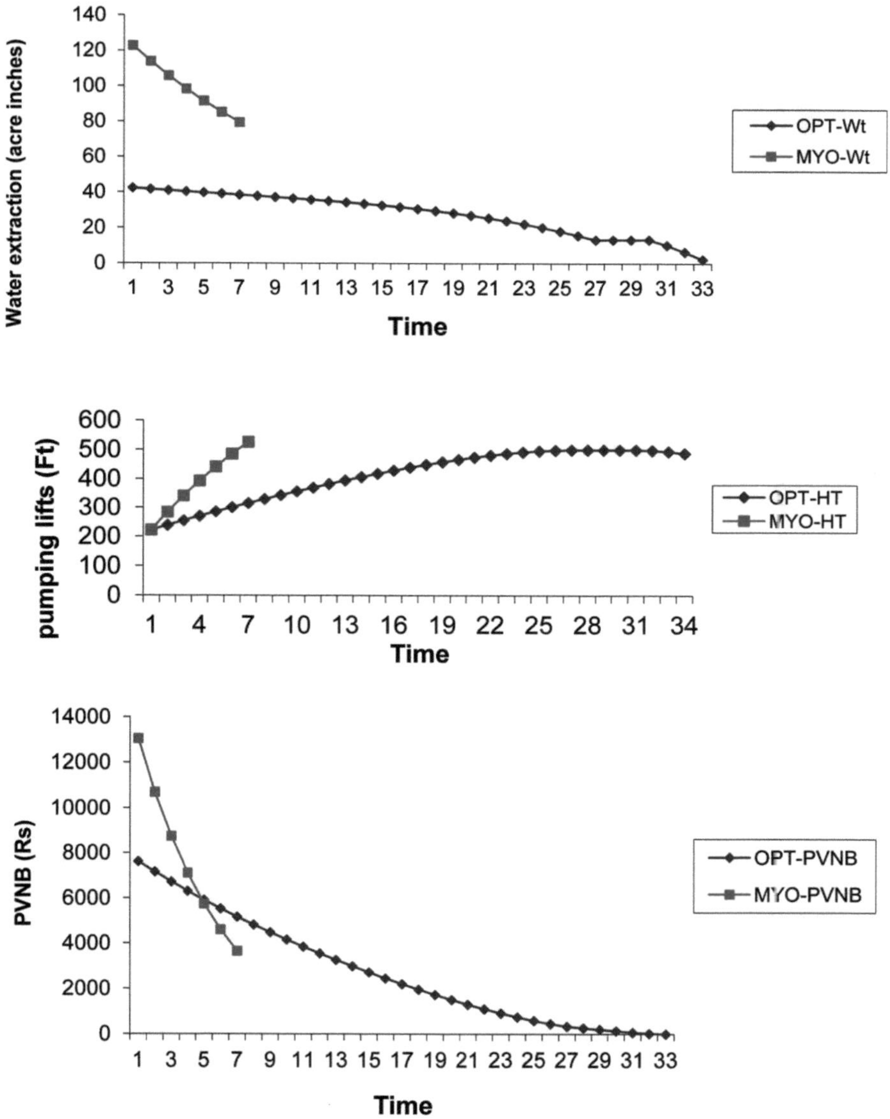

Fig. 12.2 Dug well in canal command, Saundatti taluk, Belgaum district. (Source: Rajendra 2003)

The pumping cost is

$$\mathrm{TC} = K \times h_t \times w_t,$$

where TC is the total pumping cost, K is the pumping energy cost (k_1) to lift an acre-inch of groundwater 1 in. up + the cost of negative externality (k_2).

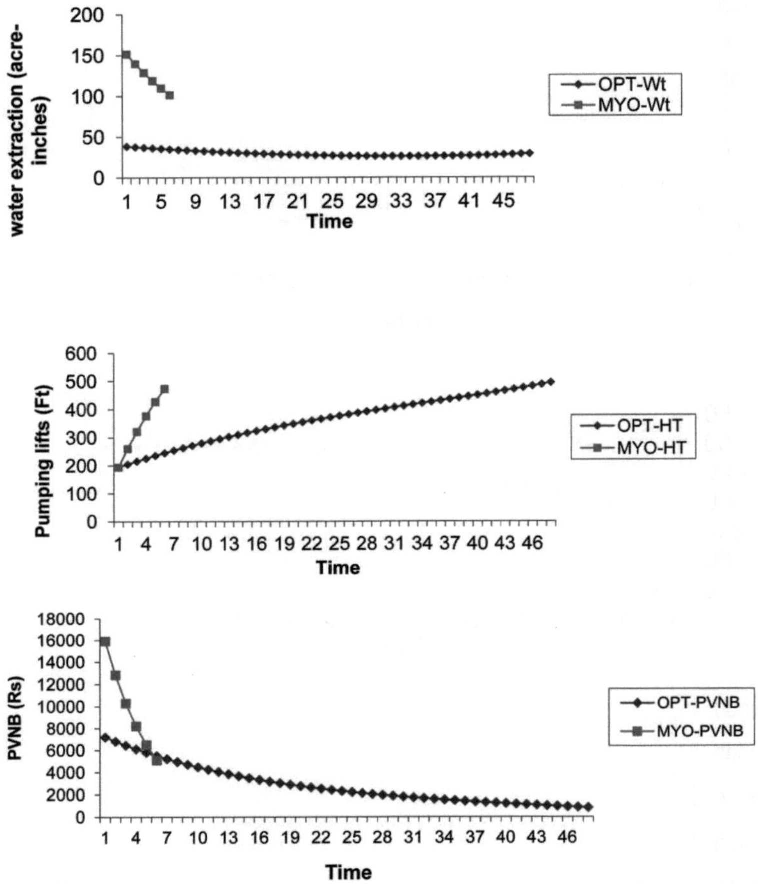

Fig. 12.3 Dug well with low water level in sole well situation, in Bailhongala taluk, Belgaum district. (Source: Rajendra 2003)

The hydrological equation is

$$h_{t+1} = h_t + \{(1-\theta)w_t - R\} / \{As\},$$

where, h_t is the initial pumping lift, which is the depth of pump placement in the borewell from the surface of the earth in inches at time t; h_{t+1} is the pumping lift in the period $t+1$, which is estimated as given in the expression; θ is the return flow coefficient of irrigation water which percolates back into the aquifer where $0 \le \theta \le 1$; R is the annual average natural recharge of the aquifer; A is the area of the aquifer: the area of the aquifer per well is equal to the geographical area divided by the number of functioning wells in the village; s is the the storativity coefficient.

Fig. 12.4 Myopic and optimal groundwater extraction, pumping lift and present value net benefits *(PVNB)* in GWSI

Fig. 12.5 Myopic and optimal groundwater extraction, pumping lift and present value net benefits *(PVNB)* in GWTI

The total revenue is the area under inverse demand curve, obtained by integrating the water demand function. Thus, the total benefit (TB) from groundwater per year is

$$\mathrm{TB}_t = \sum_{t=0}^{n} \int (a - 2bw)dw = \mathrm{TB}_t = \sum_{t=0}^{n} \left(aw - bw^2\right).$$

The NB is

$$\mathrm{NB} = \sum \left(aw_t - bw_t^2 - Kh_t w_t\right),$$

by deducting the total cost *Khw* from the TB.

Now, the objective is to

$$\mathrm{MaxNB} = \sum_{t=0}^{n} e^{-rt} \left(aw_t - bw_t^2 - Kh_t w_t\right),$$

subject to

$$h_{t+1} - h_t = \{(1-\theta)w_t - R\} / As,$$

where R is the average annual natural recharge of the aquifer.

The Hamiltonian is

$$H = e^{-rt}\left(aw_t - bw_t^2 - Kh_t w_t\right) + \lambda\{(1-\theta)w_t - R\}/As.$$

Here, λ is the marginal user cost/shadow price/royalty/opportunity cost, which is the reduction in the undiscounted future NBs from the withdrawal of one additional unit volume of water in the current period.

The three conditions for optimality are as follows:

1. $\delta H / \delta W = 0$ implies $e^{-rt}(a - 2bw_t - Kh_t) + \lambda\{(1-\theta)/(As)\} = 0$ or $e^{-rt}(a - 2bw_t) = e^{-rt}Kh_t - \lambda\{(1-\theta)/(As)\}$. Here r is the social discount rate.
2. $\delta H / \delta h = \lambda_{t+1} - \lambda_t$ implies $\lambda_{t+1} - \lambda_t = e^{-rt}Kw_t$
3. $\delta H / \delta\lambda = h_{t+1} - h_t$ implies $h_{t+1} - h_t = \{(1-\theta)W_t - R\}/\{As\}$

Thus, the objective function estimated is

$$\text{MaxNB} = \sum_{t=0}^{n} \rho^t\left(\text{TR} - \text{TC}\right),$$

subject to

$$h_{t+1} - h_t = \{1/(1+\mathrm{r})\}\{(1-\theta)W_t - R\}/\{As\},$$

with TR as the total revenue (₹ per well), TC as the total cost of irrigation (₹ per acre-inch of water per inch of lift) and ρ as the discount factor, which is $\{1/(1+r)\}$.

The total revenue is estimated by regressing the total revenue minus all costs of cultivation excepting the irrigation cost. The revenue per well is hypothesized to vary directly with water used and inversely with the square of water used with zero intercept:

$$\text{TR} = aw_t - bw_t^2.$$

Here, TR is the gross returns minus all costs of cultivation except the irrigation cost. The intercept is assumed zero to allow 'irrigation' to shape the gross returns. The total cost is $K \times ht \times w_t$. Here, $K = k_1 + k_2$, where k_1 is the electricity cost to lift an acre-inch of water 1 in. up and k_2 is the negative externality cost incurred per inch of water per inch of lift. From an earlier study conducted by Sathisha (1997) in the Department of Agricultural Economics, University of Agricultural Sciences (UAS) Bangalore, the cost of electricity required to pump groundwater is estimated. It is reported that 42 KW h energy was required to lift 1 acre-inch of water from a height of 80 ft. Therefore, the power required to lift 1 in. of water by 1 in. lift, $k_1 = 0.04375$ KW h. The power is charged at ₹ 0.50/KW h according to the recommendation of the National Council of Power Utilities. k_2 is estimated as under:

$$\text{Negative externality cost per inch of water per inch of lift}$$
$$= (\text{ACAW} - \text{ACFW}) / \text{WUF} \times (1/\text{initial pumping lift}),$$

where, ACAW is the amortized cost of all wells, ACFW is the amortized cost of functioning wells and WUF is the water used on the farm.

The recharge is estimated as $R = R_c \times A \times R_f$,

where R_c is the recharge coefficient ($0<R_c<1$), A is the area of groundwater basin per well and R_f is the rainfall (inches). The pumping lift is the distance from the earth surface to the water table, and it is given by the depth of pump placement from the base level.

Steady-state groundwater extraction (Wss) is given by

$$W\text{ss} = \frac{R}{1-\theta}.$$

The steady-state height of the groundwater level (Hss) is given by

$$H\text{ss} = -\left(\frac{R}{Asr}\right)+\left(\frac{a}{k}\right)-\left(\frac{2bW\text{ss}}{k}\right).$$

Maximizing the present value net benefit (PVNB):

$$MaxPVNB = \int_0^{\infty} e^{-rt}\left((aw-bw^2-kwh)+\lambda_t\left(\frac{(1-\theta)w-R}{As}\right)\right).$$

Myopic situation: For a myopic situation, the farmer does not worry about the opportunity cost, user rent or the royalty of groundwater. Therefore, myopic extraction levels are higher than optimal extraction limits as here the marginal revenue is equal to the marginal cost. Thus,

$$a-2bw_t = Kh_t$$
$$orw_t = B_0 - B_1h_t,$$
where
$$B_0 = a/2b$$
$$B_1 = K/2b.$$

It is assumed that the externality cost increases at 15, 7.5 and 2.5 % in GWSI, GWTI and GWCI, respectively. k_2 is estimated as follows:

Negative externality cost per inch of water per inch of lift
$= (\text{ACAW} - \text{ACFW}) / \text{TWU} \times (1/\text{initial pumping lift}).$

Here, ACAW is the amortized cost of all wells; ACFW is the amortized cost of functioning wells and TWU is the total water used.

Recharge (R): $R=R_c \times A \times R_f$. Here, R_c is the recharge coefficient ($0<R_c<1$); A is the area of groundwater basin per well and R_f is the rainfall (inches). Here, the groundwater recharge is assumed to be 5, 7.5 and 10 % in GWSI, GWTI and GWCI, respectively.

The pumping lift is the distance from land surface to water table, obtained from secondary data from the District Statistical Office, Belgaum. A close approximation to this is the average depth of pump placement from the earth. Here, the pumping lift is restricted to 500 ft. The area of the aquifer per well is approximately the total land holdings divided by the number of functioning wells.

12.2 Estimation of NB Under Myopic (No Control)

The path under myopic extraction was estimated by maximizing the NB per annum subject to the water availability and other constraints. The resulting water balance is introduced as the initial water availability for the next year and the recharge and return flows during the year were added to the initial water balance to estimate the total water availability in the current year. This procedure is repeated for all the 20 years of horizon. The individual yearwise NB was discounted and summed up to estimate the sum of the present value of NBs over the entire period. The initial input for the model is in Table 12.1.

Table 12.1 Economic and hydrological parameters of the estimated optimal control model

Sl. No.	Constants and variables	GWSI ($n=30$)	GWTI ($n=30$)	GWCI ($n=30$)
1	Aquifer area per functioning well (acres)	4.6	5.6	6.1
2	Initial pumping lift (inches)	3025	2664	2328
3	Storativity coefficient	0.025	0.025	0.025
4	Groundwater recharge (% of rainfall)	5	7.5	10
5	Groundwater recharge (acre-inches)	6.39	12.60	14.53
6	Groundwater return flow coefficient (θ)	0.05	0.05	0.10
7	k_1= cost of electrical power (₹ per acre-inch, per inch of lift)	0.0218	0.0218	0.0219
8	k_2= annual externality cost (₹ per acre-inch per inch of lift)	0.0026	0.0024	0.0025
9	Annual externality cost (Rs) assumed to increase at the rate of (%)	10.00	7.50	2.50
10	Estimated regression coefficient of groundwater extraction in quadratic function	239	284	273
11	Estimated regression coefficient of the square of groundwater extraction in quadratic function	−0.711	−0.919	−0.732
12	Discount rate chosen	0.02	0.02	0.02
13	Discount factor (=1/(1+0.02))	0.980	0.980	0.980
14	Annual rainfall (inches)	27.79	30	24

GWSI groundwater wells under sole irrigation, *GWTI* groundwater wells under tank irrigation, *GWCI* groundwater wells under canal irrigation

12.3 Intertemporal Allocation of Groundwater

The intertemporal (optimal) groundwater allocation path has been derived using the optimal control theory. The present value of NB under optimal extraction and myopic (competitive) extraction was estimated. The estimated parameters of the water revenue function $Y = aW + bW^2$ are listed in Table 12.2. Here, Y is the gross returns per well (₹), and W is the water extracted per well (acre-inches).

Path of Groundwater Use in GWSI Under Myopic Extraction If farmers follow myopic extraction of groundwater, they are not concerned about the negative externality imposed on other farmers as well as on themselves since groundwater use imposes a reciprocal externality. In this situation, the life of the well is estimated to last for 9 years. However, the restrictive assumption here is that there is no cumulative interference of irrigation wells, and there is no change in the rainfall and/or availability of groundwater. In the initial periods, extraction of groundwater will be higher, commencing from 61 acre-inch per well in the first year and falling to 39 acre-inch in the ninth year. As the discounted NBs will be low, farmer would find it economically not worthwhile to use the well from the tenth year. The total present value of NB for 9 years is ₹ 35,167 (Table 12.3).

Annual externality cost = amortized cost per functioning well − amortized cost per well = amortized cost of all wells / no. of wells

Amortized cost per functioning well
= amortized cost of all wells/no. of functioning wells

Path of Groundwater Use in GWSI Under Optimal Extraction The optimal control theory treats marginal returns as equal to opportunity cost plus the scarcity rent or externality cost. If a farmer adopts an optimal extraction path as prescribed by the control theory, the life of the well extends from 9 to 21 years. Optimal extraction conserves groundwater by extracting lower volume of water in initial periods (42 acre-inch per well) when compared with myopic extraction (61 acre-inch per well). Thus, in the initial year, the farmer has higher access to groundwater to the tune of 45 % when compared with the optimal extraction path. The water level in

Table 12.2 Estimates of quadratic water revenue function

Regime	a	b	R^2
GWSI	239 (25.6)	−0.711 (0.29)	52
GWTI	284 (30.9)	−0.919 (0.211)	47
GWCI	273 (24.8)	−0.732 (0.198)	55

Figures in brackets are standard error
GWSI groundwater wells under sole irrigation, *GWTI* groundwater wells under tank irrigation, *GWCI* groundwater wells under canal irrigation

Table 12.3 Features of irrigation wells in North Karnataka, Belgaum

Particulars	GWSI	GWTI	GWCI	All farms
Number of farmers	30 (33)	30 (33)	30 (33)	90 (100)
Number of wells	56 (40.5)	42 (30.5)	40 (29)	138 (100)
Number of functioning wells	39 (34)	37 (32)	38 (33)	114 (100)
Number of non-functioning wells	17 (71)	5 (21)	2 (8)	24 (100)
Investment on all wells compounded to year 2002 (₹ in lakhs)	2.59 (43)	19.91(28)	19.63 (28)	69.14
Investment on well irrigation per farm (₹)	98,651	66,388	65,451	72,707
Amortized cost of all wells (₹ in lakhs)	3.34 (47)	2.03 (28)	1.79 (24)	7.15 (100)
Amortized cost per well (₹)	5960	4844	4465	5127
Amortized cost per functioning well (₹)	8558	5499	4700	5756
Amortized cost per functioning well (₹)	8558	5499	4700	630
Investment per GIA	11,493	6282	5264	7298
Investment per functioning well	75,886	53,828	51,672	60,655

The study pertains to Bailhongala, Khanapur, Ramdurga and Saundatti Taluks
Figures in parentheses indicate the percentage to total
GWSI groundwater wells under sole irrigation, *GWTI* groundwater wells under tank irrigation, *GWCI* groundwater wells under canal irrigation

Table 12.4 Myopic extraction of groundwater, pumping lift and PVNBs in GWSI

Time	*Wt* (acre-inch)	*Ht* (ft)	PVNB (₹)
1	61	252	7053
2	58	290	6078
3	55	325	5046
4	52	358	4060
5	49	389	3541
6	46	418	3036
7	43	445	2511
8	41	470	2056
9	39	494	1786
			35,167

Wt Water extracted per well, *Ht* Pumping lift, *PVNB* present value net benefit

the well initially at 252 ft would be around 500 ft by the end of 21 years. The total present value of NBs for 21 years is ₹ 40,342. Thus, the optimal extraction path extends the well life by 12 years. The additional benefit due to optimal groundwater management over the myopic extraction is ₹ 5175 (Table 12.4).

Path of Groundwater Use in GWTI Under Myopic Extraction Following the myopic extraction, the life of the well is limited to 7 years. In the initial period, the farmer extracts 123 acre-inch of water, equivalent to the water requirement of 3 acres of paddy, and the corresponding PVNB is ₹ 13,043. The water level was initially at 222 ft, and by the end of the seventh year the water level would be more than two times deeper. Through myopic extraction a PVNB of ₹ 53,600 is realized in 7 years (Table 12.5).

Table 12.5 Optimal extraction of groundwater, pumping lift and PVNB in GWSI

Time	*Wt* (acre-inch)	*Ht* (ft)	PVNB (₹)
1	42	252	5664
2	41	276	5105
3	39	300	4583
4	38	322	4092
5	37	344	3630
6	35	364	3196
7	34	384	2788
8	32	403	2405
9	30	420	2046
10	28	436	1712
11	26	451	1402
12	23	464	1119
13	21	476	864
14	18	485	637
15	14	493	439
16	10	498	270
17	7	500	155
18	7	500	117
19	7	500	78
20	7	500	37
21	2	500	2
			40,342

Wt Water extracted per well, *Ht* Pumping lift, *PVNB* present value net benefit

Path of Groundwater Use in GWTI Under Optimal Extraction The extraction of groundwater according to optimal extraction extends the life of the well from 7 to 30 years. In the initial period, optimal extraction conserves water by extracting 42 acre-inch (equivalent to grow 1 acre paddy). In the 30th year, the water level stands at 500 ft when compared with myopic water level, which will be 500 ft already in the seventh year. The PVNB in optimal rule is ₹ 90,173. Thus, optimal extraction enhances the well life by about 23 years, and the additional PVNB by considering user cost is ₹ 36,573 (Table 12.6).

Path of Groundwater Use in GWCI Under Myopic Extraction If the farmer adopts myopic extraction, the life of irrigation wells is 6 years. In the initial period, the water level which stood at 194 ft, farmer reaps ₹ 15,920 of the NBs by extracting 152 acre-inch of water per well. Gradually, the water level draws down, and in the sixth year the depth of water will be around 473 ft. Compared to the sole well regime, canal command farmers initially have additional access to groundwater to the tune of 149 %, and the additional PVNB for GWCI farmers is 146 %. The total PVNB for the period of 6 years is ₹ 58,877 (Table 12.7).

Path of Groundwater Use in GWCI Under Optimal Extraction Optimal extraction extends the life of the well to 48 years, compared with myopic extraction of 6

Table 12.6 Myopic extraction of groundwater, pumping lift and PVNB in GWTI

Time	*Wt* (acre-inch)	*Ht* (ft)	PVNB (₹)
1	123	222	13,043
2	114	284	10,702
3	106	341	8746
4	98	393	7112
5	91	441	5745
6	85	486	4603
7	79	526	3648
			53,600

Wt Water extracted per well, *Ht* Pumping lift, *PVNB* present value net benefit

Table 12.7 Optimal extraction of groundwater, pumping lift and PVNB in GWTI

Time	*Wt* (acre-inch)	*Ht* (ft)	PVNB (₹)	Time	*Wt* (acre-inch)	*Ht* (ft)	PVNB (₹)
1	42	222	7623	17	31	439	2197
2	42	238	7161	18	29	449	1956
3	41	254	6724	19	28	458	1725
4	40	270	6308	20	27	467	1505
5	40	285	5910	21	25	474	1296
6	39	300	5530	22	24	481	1099
7	38	315	5167	23	22	487	915
8	38	329	4818	24	20	492	746
9	37	343	4482	25	18	496	592
10	36	357	4159	26	16	499	453
11	36	370	3848	27	13	500	335
12	35	382	3548	28	13	500	271
13	34	395	3259	29	13	500	205
14	33	407	2979	30	13	500	137
15	33	418	2709	31	10	500	68
16	32	429	2448				90,173

Wt Water extracted per well, *Ht* Pumping lift, *PVNB* present value net benefit

years. Initially, the extraction is lower and hence the PVNB too. After following the optimal path, the life of the well is extended and the farmer reaps additional benefit of ₹ 77,020 (Table 12.8).

Path of Groundwater Use in GWSI Under Myopic and Optimal Extraction If the farmer chooses myopic extraction, it exerts pressure on the aquifer to extract higher volumes of water to the tune of 45 % compared to optimal extraction in the very first year of well extraction. If every farmer resorts to this, as has been happening in some of the dry agroclimatic zones of Karnataka, it will lead to cumulative interference and large-scale initial failures. Even though water extraction in myopic rule over optimal control is higher by 45 %, the PVNB in myopic over optimal extraction is higher by only 15 %. Thus, the benefits in myopic rule are not com-

Table 12.8 Myopic extraction of groundwater, pumping lift and PVNB in GWCI

Time	*Wt* (acre-inch)	*Ht* (ft)	PVNB (₹)
1	152	194	15,920
2	140	261	12,837
3	129	321	10,302
4	119	377	8218
5	110	427	6505
6	102	473	5096
			58,877

Wt Water extracted per well, *Ht* Pumping lift, *PVNB* present value net benefit

Fig. 12.6 Myopic and optimal water extraction, pumping lift and present value net benefits *(PVNB)* in GWCI

mensurate with the volume of water extracted. Thus, the consideration of user cost extends the life of the well by about 12 years and also realizes an additional PVNB of ₹ 5175 (Fig. 12.6; Table 12.9).

Path of Groundwater Use in GWTI Under Myopic and Optimal Extraction The concern for wise use of groundwater is crucial when the balance between demand and supply for groundwater in water-scarce areas is disturbed, resulting in the decline of the groundwater table. Efficiency comes into action when water level drops; externality cost increase; cumulative well inference; increase in area under water-intensive crops, such as sugarcane, maize, cotton, paddy, etc., with the use of high-power lifts from deep water table; negligence of traditional water sources, such as tanks, etc., have all led to overextraction of water. Thus, optimal groundwater extraction is a strategy recommended for water-scarce regions to conserve

Table 12.9 Optimal extraction of groundwater, pumping lift and PVNB in GWCI

Time	*Wt* (acre-inch)	*Ht* (ft)	PVNB (₹)	Time	*Wt* (acre-inch)	*Ht* (ft)	PVNB (₹)
1	38	194	7194	26	27	382	2122
2	38	205	6808	27	26	387	2034
3	37	215	6447	28	26	392	1950
4	36	226	6109	29	26	397	1871
5	35	235	5791	30	26	402	1795
6	35	245	5493	31	26	407	1723
7	34	254	5212	32	26	412	1654
8	33	263	4948	33	26	417	1589
9	33	271	4700	34	26	421	1526
10	32	280	4467	35	26	426	1466
11	32	288	4248	36	26	431	1408
12	31	295	4041	37	26	436	1353
13	31	303	3846	38	26	440	1300
14	30	310	3663	39	26	445	1248
15	30	317	3491	40	26	450	1198
16	30	324	3328	41	27	455	1149
17	29	330	3174	42	27	460	1102
18	29	337	3030	43	27	466	1055
19	28	343	2893	44	27	471	1009
20	28	349	2764	45	28	476	964
21	28	355	2642	46	28	482	918
22	27	360	2526	47	28	488	873
23	27	366	2417	48	29	494	827
24	27	372	2313				135,897
25	27	377	2215				

Wt Water extracted per well, *Ht* Pumping lift, *PVNB* present value net benefit

and maximize returns per unit of water, and it should not result in proliferation of irrigation wells. In this direction, technological alternatives such as sprinkler and drip irrigation need to be examined to rescue farmers from water problems. Though optimal extraction is frequently suggested, it is not practical as its implementation is a challenging task (Fig. 12.7).

Path of Groundwater Use in GWCI Under Myopic and Optimal Extraction In myopic rule, the farmer reaps the benefit of groundwater extraction in 6 years. He/she has no compelling reason to drill new wells after the failure of the first well. Thus, farmers are forced to invest in a new well after the sixth year or deepen the well or change the water use technology, which are the backstop technology for the farmer to remain on the original isorevenue curve. Moreover, given the importance of farming, farmers cannot be expected to shift from farming. Investment, on water use efficiency rather than on new wells has to be encouraged and such a strategy of saving water should not reduce the income and employment. However, drip and sprinklers are rarely in use. Popularisation of such strategies enhances the water level without shifting to a lower isorevenue curve (Fig. 12.8).

Fig. 12.7 Dug well with water level, indicating synergistic effect of canal in Ramdurga taluk, Belgaum district. (Source: Rajendra 2003)

Fig. 12.8 Failed open well in sole well situation, Bailhongala taluk, Belgaum district. (Source: Rajendra 2003)

12.4 Implications

In this chapter, the results of optimal control theory applied to groundwater in tank command, canal command and well command scenarios undertaken in Bailhongala, Saundatti, Ramdurga and Khanapur taluks of Belgaum district, Karnataka indicate the following: In the areas where the access to water for irrigation is limited, and the access to groundwater irrigation is a function of recharge, there are opportunities for optimal extraction for augmenting the groundwater resource for irrigation. The major crops cultivated are cotton, jowar, sunflower, groundnut, maize and vegetables in GWSI. In GWTI and GWCI, cotton, jowar, sunflower and water-intensive crops such as sugarcane and paddy were cultivated. The amortized cost per acre of gross groundwater-irrigated area (GGWIA) was lower to the extent of 49 % (₹ 654) and 39 % (₹ 786) for GWCI and GWTI farmers, respectively, as compared to GWSI farmers (₹ 1296). To obtain 1 acre-inch of water, GWCI farmers incurred 65 % lower cost (₹ 43) when compared with GWSI farmers (₹ 123). The GWCI farmers could realize 15 % additional net return per acre-inch of groundwater compared with GWSI. GWCI and GWTI farmers have, respectively, 183 % and 110 % higher access to water for every rupee of investment than GWSI farmers. This is a prima facie evidence of groundwater recharge in the command areas. Amortized cost per functioning well was ₹ 8558, 5499 and 4700 in GWSI, GWTI and GWCI, respectively. Optimal path of extraction of the groundwater resource enhances the well life by additional 12, 24 and 42 years in GWSI, GWTI and GWCI, respectively, over the myopic extraction. The additional net present value of benefit realized is ₹ 5175, 36,573 and 77,020 by optimal extraction in GWSI, GWTI and GWCI, respectively. Thus, optimal extraction in the form of water-saving technologies such as drip irrigation promotes the sustainability of groundwater extraction as well as dampens the groundwater cost.

References

Rajendra A (2003) Optimal extraction of groundwater resource in canal, tank and well irrigation commands in Karnataka- an application of control theory. M. Sc. (Agri) Thesis (unpublished), University of Agricultural Sciences, Bangalore

Sathisha KM (1997) resource economics study of valuation of well interference externalities in central dry zone of Karnataka. M. Sc. (Agri.) Thesis (Unpublished), Department of Agricultural Economics, UAS, Bangalore

Chapter 13
Water Policy

In India, policy with regard to food, fertilizers and credit have been receiving the highest priority at both the state and union government level. Among the natural resources, the highest attention has been accorded to forest policy through the enactment of the Forest Conservation Act of 1980 and the National Forest Policy of 1988. The national water policy outlined by the Ministry of Water Resources, Government of India highlights the policy in 2012 and in 2002. Water policy of 2012 has been built on a comprehensive and action-oriented plan highlighted in the 2002 policy. Both the policy documents adequately highlight the problem of scarcity of surface water and groundwater, overexploitation, inefficient use, need for introducing improved technologies of water use, including incentivizing efficient water use, encouraging community-based management of aquifers and artificial recharging projects. The policy also emphasizes on integrated water resources management (IWRM) considering the needs of both surface water and groundwater and all uses and users in order to enhance land and water productivity. The policy has also emphasized on declining groundwater levels in overexploited, areas focusing on the role of technology and incentives. However, little has been achieved in terms of sustainable use in irrigation.

Section 4.3 focuses on demand side management: "The adaptation strategies could also include better demand management, particularly, through adoption of compatible agricultural strategies and cropping patterns and improved water application methods, such as land leveling and/or drip/sprinkler irrigation as they enhance the water use efficiency, as also, the capability for dealing with increased variability because of climate change. Similarly, industrial processes should be made more water efficient."

Section 5.4 emphasizes on falling groundwater levels: "Declining ground water levels in over-exploited areas need to be arrested by introducing improved technologies of water use, incentivizing efficient water use and encouraging community based management of aquifers. In addition, where necessary, artificial recharging projects should be undertaken so that extraction is less than the recharge. This would allow the aquifers to provide base flows to the surface system, and maintain ecology."

M.G. Chandrakanth, *Water Resource Economics*, DOI 10.1007/978-81-322-2479-2_13

Section 5.6 suggests: "Integrated Watershed development activities with groundwater perspectives need to be taken in a comprehensive manner to increase soil moisture, reduce sediment yield and increase overall land and water productivity. To the extent possible, existing programs like MGNREGA may be used by farmers to harvest rain water using farm ponds and other soil and water conservation measures."

Further in Sect. 6.5 while discussing adaptation to climate change, the demand side is emphasized but not the supply side: "Water saving in irrigation use is of paramount importance. Methods like aligning cropping pattern with natural resource endowments, micro irrigation (drip, sprinkler, etc.), automated irrigation operation, evaporation-transpiration reduction, etc., should be encouraged and incentivized. Recycling of canal seepage water through conjunctive ground water use may also be considered." There is a separate Sect. 6 on "Demand Management and Water Use Efficiency" with Sect. 6.3 stating that "Recycle and reuse of water, including return flows, should be the general norm" and Sect. 6.5 highlighting that "Water saving in irrigation use is of paramount importance. Methods like aligning cropping pattern with natural resource endowments, micro irrigation (drip, sprinkler, etc.), automated irrigation operation, evaporation-transpiration reduction, etc., should be encouraged and incentivized. Recycling of canal seepage water through conjunctive ground water use may also be considered."

Section 6.6 highlights that the "Use of very small local level irrigation through small bunds, field ponds, agricultural and engineering methods and practices for watershed development, etc. need to be encouraged. However, their externalities, both positive and negative, like reduction of sediments and reduction of water availability, downstream, may be kept in view." Section 6.7 highlights the role of water users in monitoring water use pattern: "There should be concurrent mechanism involving users for monitoring if the water use pattern is causing problems like unacceptable depletion or building up of ground waters, salinity, alkalinity or similar quality problems, etc. with a view to planning appropriate interventions."

Section 7 on water pricing indicates that water pricing should reflect efficient use of water as well as rewarding water conservation efforts, as in Sect. 7.1: "Pricing of water should ensure its efficient use and reward conservation. Equitable access to water for all and its fair pricing, for drinking and other uses such as sanitation, agricultural and industrial, should be arrived at through independent statutory Water Regulatory Authority, set up by each State, after wide ranging consultation with all stakeholders." Further, in Sect. 7.2, the role of equity and efficiency is highlighted charging for surface water including the need for volumetric measurement of water: "In order to meet equity, efficiency and economic principles, the water charges should preferably/as a rule be determined on volumetric basis. Such charges should be reviewed periodically." There has been no effort in the States, either by the Department of Agriculture/Horticulture or Water resources Departments towards volumetric measurement at farm level. This has been the biggest flaw pertaining to water measurements constraining water budgeting. When water is not measured volumetrically, the question of charging for water weakens. Equity is further emphasized in Sect. 7.4, indicating the role of differential pricing for uses of water ensuring food and livelihood security of the poor: "The principle of differential pricing may be retained for the pre-emptive uses of water for drinking and sanitation; and high priority allocation for ensuring food security and supporting livelihood for the poor. Available water, after meeting the above needs, should increasingly be subjected to allocation and pricing on economic principles so that water is not wasted in unnecessary uses and could be utilized more gainfully."

The role of water users associations (WUAs) is highlighted in Sect. 7.5 in charging for water and in volumetric measurement of water: "Water Users Associations (WUAs) should be given statutory powers to collect and retain a portion of water charges, manage the volumetric quantum of water allotted to them and maintain the distribution system in their jurisdiction. WUAs should be given the freedom to fix rates subject to floor rates determined by Water Resources Authority (WRA)." However, the role of WUAs is limited to surface water and does not include groundwater.

Section 7.6 highlights that groundwater use be regulated by regulating the use of electricity and not through the awareness of farmers, creating incentive structures for efficient use: "The over-drawal of groundwater should be minimized by regulating the use of electricity for its extraction. Separate electric feeders for pumping ground water for agricultural use should be considered."

Section 12.4 mentions institutional arrangements emphasizing on IWRM: "Integrated Water Resources Management (IWRM) taking river basin/sub-basin as a unit should be the main principle for planning, development and management of water resources. The departments/organizations at Centre/State Governments levels should be restructured and made multi-disciplinary accordingly."

Finally, Sect. 16.1 suggests the constitution of a national water board to prepare a plan of action for implementing national water policy: "National Water Board should prepare a plan of action based on the National Water Policy, as approved by the National Water Resources Council, and to regularly monitor its implementation." Since water is a state subject, the national water policy suggests that the states "may" draft/revise their policy in accordance with the national water policy: "16.2 The State Water Policies may need to be drafted/revised in accordance with this policy keeping in mind the basic concerns and principles as also a unified national perspective."

13.1 Weak Points of Water Policy

1. Emphasis on the demand side: There is an emphasis on the "demand side" rather than "supply side" management. For instance, there are incentives for "drip irrigation", while there are no incentives for "groundwater recharging of borewells".
2. Emphasis on the use of electricity as a lever for groundwater regulation: It is important to note that, it is not the "electricity" but the "groundwater" resource which is crucial. However, the researchers and policy makers focus on electricity subsidy for groundwater farmers, rather than educating the farmers or their capacity building pertaining "water budgeting", sustainable water use and con-

servation of water including recharge. By largely focusing on electricity subsidy, farmers are given a feeling that they are receiving substantial gains. However, through proper costing of irrigation wells, including initial and premature failure of wells, considering their life and age and through crude volumetric measurement of groundwater pumped, farmers need to be educated that the electricity they are receiving forms the tip of the iceberg in the cost of groundwater, since the farmer bears at least 70 % of the cost of groundwater through investment on irrigation wells, pump sets, pump house, conveyance structures, drip irrigation investment if any, while 30 % may amount to the cost of electricity subsidy. It is necessary that an estimate of volumetric measurement of groundwater is made by farmers and they are educated regarding this. Thus, low-cost water measuring equipment including low-cost water meter and low-cost water level indicator needs to be fixed on borewells rather than fixing "electrical meters". Fixing water meters will also not invite the wrath of farmers as water measurements are helpful in water budgeting. Certainly, after farmers are convinced, through capacity-building programmes, they can be asked to bear some proportion of pumping costs of electricity.

3. Poor calculation of the cost of groundwater irrigation in the cost of cultivation: At present, the cost of cultivation surveys of the state (farm management surveys) or the cost of cultivation schemes sponsored by the Directorate of Economics and Statistics all over the country do not calculate the cost of groundwater or surface water properly. It is necessary that the cost of cultivation studies of the governments include a proper method to first measure the water applied for irrigated crops and the cost of water used in irrigation. Because the water is neither volumetrically measured nor is the cost calculated, the profits from irrigated crops are "pseudo profits levels" which are overestimated. By proper cost calculation for water, the government will be able to obtain the proper estimate of the cost of cultivation, which at present does not include the cost of water/groundwater. This will enable the government/s to fix the minimum support price (MSP) for irrigated crops, which should also differentiate between surface-water-irrigated and groundwater-irrigated crops as the cost of surface water is below 5–10 % of the cost of groundwater.
4. Absence of irrigation extension: At present, there is no programme of "irrigation extension" similar to "agricultural extension" as the farmers/water users are not aware of the complexity of using and conserving water resources including payment for the use of water. Agriculture is the largest user of water. At present, out of the total water use, irrigation is the largest user of water resources. And groundwater meets 80 % of the irrigation needs. The water policy does not adequately address how vital/crucial groundwater is when compared with surface water. Groundwater has become an orphan when compared with the focus and attention which the surface water is receiving, which only meets 20 % of the nation's needs. In addition, groundwater investments are made by farmers as private investments, while surface water investments are made from public funding. The public investment in groundwater happens largely through "electricity subsidy", which forms the "tip of the iceberg" of groundwater cost. The

policy suggests "regulation" of groundwater through "electricity pricing", given the fact that farmers meet at least around 70 % of the cost of groundwater, while the government meets virtually 100 % of the cost of surface water.

5. Capacity-building programmes: Farmers need to be educated regarding the proper use of water, to reduce externalities in the use of water such as salinity, alkalinity and water logging. This requires creation of "irrigation management service" involving agriculture graduates/agricultural engineering graduates/horticulture graduates, who can play a major role in creating awareness as well as in monitoring the use of the scarce water resources.
6. Lack of focus of policy: Water policy has not received sufficient focus and attention as the "National Forest Policy" of 1988 and the "Forest Conservation Act" of 1980. One of the reasons is that forest comes under concurrent list, while water is under the state list in the constitution. Water should also be included in the concurrent list of the constitution. In fact, Sect. 16.2 says that the states "may" redraft their water policy in line with the national water policy, which also means that the states "may not" comply, and there is no hold by the centre regarding the compliance. However, this does not happen in the case of forest policy.
7. Poor reflection of groundwater overexploitation: Policies such as "Million wells scheme" and "Ganga Kalyana Yojana" were based largely on the hypothesis that groundwater has not been exploited to the extent that it could not serve the food and livelihood needs anymore. However, such policies were lopsided, as they were more focused on the "demand side" relegating the "supply side". The policy was not adequately backed up by how to make those wells sustainable; with the result that majority of the borewells suffered from well failure and resulted in abandoning the wells with high social cost of loss of public investment. Unless the "demand side" is adequately balanced by the "supply side" of water resources, inefficient and unsustainable use is bound to occur and recur in future.
8. Water in the "state list": As water continues to be in the state list, and the state does not treat "water" as seriously as required, water policy is relegated. In states, groundwater is clubbed with mines, and hence mines receive greater importance than groundwater, as minerals are expensive and are of commercial value. It is crucial to handle groundwater and water through a separate department, which should involve hydrogeologists, engineers, agricultural graduates and agricultural engineering graduates.

The water policy thus needs to be strengthened through (1) governance structure (2) accountability (who has to do what and what are the incentives/penalties for compliance/noncompliance) and (3) focusing on the supply side.

Chapter 14
Estimated Cost of Surface Water and Payment for Ecosystem Services (PES) for Water

14.1 Culture of Sacred Groves

The practice of sacred groves of Kodagu where farmers/planters worship the nature through trees is linked with their devotion to river Cauvery. The culture of farmers in their ancestral worship through reverence to Iyenmane (the house of the ancestor or the 'Karona'), the ancestral burial place or 'kaimada', the trees around the ancestral burial place or 'karona kadu', the reverence to the sacred grove or 'devara kadu', which has old-growth trees facilitating nature worship, a sacred site and a sacred tank have been part of the forest ecosystem cultivating coffee, cardamom and pepper in Kodagu, in the Western Ghats recognized as the hottest of the hot spots of biodiversity by the International Union for Conservation of Nature (http://ribm.net/kodagu-model-forest). In this regard, the role played by the Kodagu Zilla Devakad Takkamukyasthara Mahamandali, Madikeri and the College of Forestry, Ponnampet, University of Agricultural and Horticultural Sciences (UAHS), Shimoga is laudable.

14.1.1 How PES Is Linked with Farmers' Efforts in the Cauvery Basin

The Cauvery basin is unique compared to other river basins in India, where more than 97 % of the river water flow is put to use in agriculture as irrigation, domestic use and industrial use. The hydrological services rendered by the catchment areas in Brahmagiri and the upper portion of the Cauvery catchment contribute to the river flow which is complemented by the efforts of the farmers and planters of Kodagu, who are maintaining the biodiversity in coffee estates through shade coffee facilitated by the culture of sacred groves. Therefore, it is in order that such conservation efforts of the farmers need to be supported and paid for by the users of the Cauvery water in the downstream of the basin. This chapter is a modest attempt in this endeavour (Chandrakanth and Nagaraja 2014).

M.G. Chandrakanth, *Water Resource Economics*, DOI 10.1007/978-81-322-2479-2_14

14.2 Cost of Water

Surface water is usually from an irrigation reservoir/dam/tank with runoff from rainfall. In order to obtain the fixed cost of water, time series data on investment on construction of an irrigation reservoir/dam and canals need to be obtained along with operation and maintenance costs. The annual cost data or cash outflows are compounded to the present at a real social discount rate (of around 2 %), which reflect the total cost of irrigation structure assuming that it can be built in the latest 1 year. The compounded total cost is amortized over, say, 100 years, assumed to the life of the irrigation reservoir. The amortized cost which gives the cost per year is divided by the gross area irrigated and reflects the value of irrigation water per acre. This exercise for an irrigation dam in Karnataka worked to around ₹ 1000/acre (Nagaraj et al. 2003). The social discount rate of 2 % reflects the criterion of sustainability and intra- and intergenerational equity. Sugarcane and paddy are commonly cultivated crops in canal irrigation in the Cauvery basin of Karnataka. It is crucial to note that the hydrological services of the Cauvery River are due to the watershed practices followed by the farmers of the upper Cauvery basin located in Kodagu, because of which the river water flows without much of the sedimentation.

Supposing that crops are irrigated by borewells on volumetric basis, the cost works to ₹ 500/acre-inch in dry agroclimatic zones (Chandrakanth 2009). With the consumptive use of 40 acre-inch for paddy, the cost of groundwater works to ₹ 20,000/acre. For sugarcane, the consumptive use is 70 acre-inch of water, and the cost works to ₹ 35,000/acre. While surface water is an explicit benefit amenable for ecosystem service, groundwater is an implicit benefit which needs to be accounted. The Revenue Department in Karnataka charges a water rate of ₹ 100/acre for paddy, ₹ 400/acre for sugarcane and ₹ 66/acre for semidry crops (jowar, groundnut, tur). However, due to a large number of small and scattered holdings, the transaction cost of collecting water rate by the Water Users Cooperative (or Association) or by the Revenue Department is colossal. Also, due to political economy, governments shy away from asking farmers to pay for water for irrigation. Therefore, irrigation charges are seldom collected by the state and seldom paid by the farmers.

14.3 PES for Water

The yield of paddy is around 20 quintals and that of sugarcane is around 600 quintals/acre. Most of the paddy cultivated is from surface water and similarly sugarcane. When the water cost of ₹ 1000/acre of irrigated land is measured on crop output, the surface water cost works to ₹ 50/quintal of paddy and ₹ 1.67/quintal of sugarcane. Paddy is milled to rice in rice mills. At least 50 % of sugarcane is crushed to sugar in sugar mills (while the balance goes to jaggery making). Considering water cost of ₹ 50/quintal of paddy and ₹ 1.67/quintal of sugarcane, rice mills can collect Rs 50/ quintal of paddy processed, and sugar mills can collect ₹ 1.67/quintal of sugarcane processed as PES. With the conversion of paddy to rice being 60 %, this amounts to ₹ 50 for 60 kg of rice, and the PES works to 83 paise/kg of rice. Similarly, considering the sugar recovery of 10 %, the PES of ₹ 1.67/quintal of sugarcane amounts to ₹ 1.67 for 10 kg of sugar or 17 paise/kg of sugar. Thus, for rice mills and sugar mills, collecting 83 paise/kg of rice or 17 paise/kg of sugar will not be a burden and will entail low transaction cost. As the number of rice mills and sugar mills is small, there will be accountability in collecting processing fee with negligible transaction cost if the payment vehicle is robust. The PES in both paddy and sugarcane is ultimately transferred to farmer beneficiaries, including consumers.

14.4 Payment Vehicle

In PES, the payment vehicle (which is the mode through which payment is sought or generated) is crucial as transaction cost of collection should be low and accountable. Surface irrigation water is mainly used for cultivating paddy and sugarcane. With the advent of rice mills and sugar mills, processing paddy in rice mills and sugarcane in sugar mills is indispensable. Hence, processing fee on rice and sugar reflecting the use value of surface water can be an effective robust payment vehicle. The processing fee can also be collected as a fee in the electricity bill by the state. Since the collection of electricity fee through electricity bill is easy, the transaction cost of collecting the PES will be negligible. Finally, the PES is transferred to beneficiaries/consumers on per kilogram basis, but it generates sizeable revenue for the state to honour the conservation efforts of the farmers.

The PES thus generated can be shared with the farmers/planters responsible for the flow of clean water downstream. A part of the PES funds can be used for capacity building regarding economic use of water and water budgeting. Water flow on the Indian peninsula is largely from rainfall. Soil and water conservation, watershed management and agro forestry efforts in upstream contribute to the water flow in downstream. Economic valuation of ecosystem services is crucial to value conservation efforts. Cultivation practices in coffee estates in Kodagu, Karnataka, inter alia soil and water conservation measures, the conservation of Devara Kadu sacred groves, the management of shade in coffee plantations using a wise combination of tree species and the maintenance of topography influence quality and volume of water flow in the Cauvery basin. A part of the PES collected through electricity fee from rice mills and sugar mills can be shared with the farmers in Kodagu as a reward for their conservation efforts in contributing to the flow of clean water in the downstream Cauvery basin.

References

Chandrakanth MG (2009) Karnataka State water sector reform: current status, emerging issues and needed strategies. International Water Management Institute, Hyderabad. http://toenre.com/downloads.html

Chandrakanth MG, Nagaraja MG (2014) Payment for ecosystem services for water—Case of Cauvery. Current Science 107(9):1375–1376 (http://www.currentscience.ac.in/Volumes/107/09/1375.pdf)

Nagaraj N, Shankar K, Chandrakanth MG (2003) Pricing of irrigation water in Cauvery Basin–Case of Kabini Command. Economic and Political Weekly 38(43):4518–4520

Chapter 15
Economics of Artificial Recharge of Borewell in Hard-Rock Areas

15.1 Recharge of Borewells

Even though studies in the Department of Agricultural Economics at the University of Agricultural Sciences, Bangalore focused on recharge of groundwater, most of them concentrated on 'macro' efforts to recharge such as the economic impact of watershed development programs (Chandrakanth et al. 2004; Seema et al. 2008, 2009), rehabilitation of irrigation tanks, soil and water conservation, joint forest planning and management of recharge (Patil et al. 2014). During the times when such studies were conducted, there were few or no attempts towards recharging individual borewells on the farms. In the recent years, technology of individual borewell recharge has been emerging, and farmers are adopting this technology gradually. The progress is certainly sluggish. Obviously, earlier studies could not focus on the economics of recharging borewells on farms. This chapter is a modest attempt towards highlighting the economic benefits of recharging individual borewells on farms.

Therefore, even though the supply side efforts to augment water, such as watershed development projects, tank rehabilitation, joint forest planning and management, have been a part of the non-plan expenditure of the government, individual borewell recharge efforts have not received the attention of policymakers and farmers as the literacy towards such an innovative endeavour as well as objectivity in analysis is still in infancy. For example, it is difficult to convince any farmer to take up recharging his/her borewell since the very first question posed is, 'How can my well be recharged since it is very deep, and even if I do the recharge, what is the guarantee that my recharge effort will only benefit my well and not the neighbouring well?' Such questions are not only difficult to be answered, but also there are implicit rather than explicit answers.

The economic importance of individual borewell recharge efforts by farmers is yet to catch the attention of researchers, policymakers and farmers. This may be due to lack of trust and information on the potential benefits of borewell recharge. In this regard, the efforts of the hydrogeologist, NJ Devaraj Reddy, Chitradurga district,

M.G. Chandrakanth, *Water Resource Economics,* DOI 10.1007/978-81-322-2479-2_15

and Progressive Farmer Sri Anjaneya Reddy in Chikkaballapur—Kolar Districts who have taken this task on a mission mode, are laudable. Obviously, such efforts are endemic and are in a niche. This study reports economics of an individual borewell recharge effort in Karnataka using partial budgeting technique (Patil 2014).

15.2 Economic Benefit of Recharge Using Partial Budgeting

Partial budgeting measures the economic impact of introducing partial changes on the farm and highlights the additional benefits and expenditures due to such changes. The four components of partial budgeting are as follows: First, the *added costs* due to the new technology of borewell recharge in relation to control; they include all increased expenses due to the new technology over control. The second component is *reduced returns* or reduced income due to the new technology of borewell recharge. Both these add to the 'costs' or 'debit' side in the partial budget. Third, the *reduced costs* due to the new recharge technique are the most crucial aspect of partial budgeting. Basically, it refers to the savings as a result of the new technology. Finally, there are *added returns* due to the new technology of recharge in terms of increase in yield, fodder, quality and other benefits. The reduced costs and added returns together form the credit side of the partial budget. If the credit is more than the debit side, the new technology of borewell recharge is economically viable. Each component is further explained below.

15.3 Added Costs

These include the labour charges for digging a 10 ft^3 pit around the borewell, boulders of different sizes, sand, charcoal, drilling holes and covering them by nylon mesh. This may amount to ₹ 25,000/well. Considering that the recharge structure serves for 10 years, the investment per year amounts to ₹ 2500/well (Table 15.1). This can also be amortized at 2% and considered.

15.4 Reduced Returns

When there are no attempts to recharge, there are frequent failures of borewells leading to reduction in the yield of groundwater. Hence the irrigated area gets reduced as the water yield cannot irrigate all the area. Thus, reduced returns include the gross return loss per recharged well when there is no recharge. The well serves 4 acres. The lost return is equal to the gross return from the functioning well on a farm with no borewell recharge, which is equal to around ₹ 400,000 for artificial

Table 15.1 Economic impact of artificial recharge of borewell (₹/borewell)

Added costs of borewell recharge	Reduced costs of borewell recharge
Amortized cost of groundwater recharge per well year = 2500/well/year	Reduced cost of borewell recharge is the difference in externality equal to the externality on the recharge farm over non-recharge farm. The externality per borewell on the recharge farm is ₹ 3000 and on farms with no recharge is ₹ 10,000. The reduced externality due to recharge is 7000
Reduced returns due to borewell recharge	*Added returns of borewell recharge*
Gross return foregone due to borewell recharging = 100,000/acre × 0.00075 acre = ₹ 75 considering the area devoted for groundwater recharge around the well = 0.00075 acre	Farmers realize gross return of ₹ 550,000/ recharged borewell
In addition, if there is no recharge, some cultivated area will be lost, as the farmer cannot irrigate his/her area. Thus, the reduced return due to no or low recharge = return realized on farm with no borewell recharge or control farms. The gross return per functioning well on the control farm = ₹ 400,000	
Debit side = 2500 + 400,000 + 75 = ₹ 402,575	Total credit side = ₹ 557,000
Credit minus debit = 557,000–402,575 = 154,425	

recharging of borewells, when 10 ft^3 area is dug around the borewell. The income lost due to this area devoted to recharge (= 0.000075 acre) × the gross return/acre = ₹ 400,000/4 × 0.000075 acre = ₹ 75. Thus, the total of debit side due to groundwater recharge is ₹ 2500 + 75 + 400,000 = ₹ 402,575.

15.5 Reduced Costs

Due to the recharge of a borewell, the externality per well gets reduced compared with a borewell which is not recharged. Thus, the negative externality per borewell with no recharge is 10,000, while the negative externality on a farm with recharged borewell is ₹ 3000. Thus, the reduced externality = ₹ 7000/borewell.

15.6 Added Returns

The gross return per recharged borewell is ₹ 550,000 and hence the credit side is 7000 + 55,000 = 557,000.

The difference between the credit side (₹ 557,000) and debit side (₹ 402,575)=₹ 154,425 indicates the economic worth of borewell recharge (Table 15.1).

15.7 Conclusion

It is economically worthwhile to recharge a borewell as the benefits outweigh the costs substantially. The cost of borewell recharge does not form even around 10 % of the cost of a new borewell but has substantial returns in the short run. The long-run impact of borewell recharge leads to sustainability in recharge. It is crucial to note that such recharging of borewells is possible and plausible even for deep borewells, as there is still a sort of doubt in the minds of farmers and scientists, as to how deep borewells can be recharged.

References

Chandrakanth MG, Alemu B, Bhat MG (2004) Combating negative externalities of drought groundwater recharge through Watershed Development Programme. Econ Politic Wkly 39(11):1164–1170. http://www.toenre.com/downloads/2004_03_combating_neg_externalities_MGC_Bisrat_EPW.pdf. Accessed 13 March 2004

Patil KKR (2014) Economics of coping mechanisms in groundwater irrigation: role of markets, institutions and technologies. Unpublished PhD thesis, submitted to the Department of Agricultural Economics, University of Agricultural Sciences, Bangalore

Patil V, Chandrakanth MG, Gangadharappa NR, Manjunatha AV, Shivanagouda B (2014) Economic impact of forest management institutions on groundwater recharge in Karnataka, India. Indian For 140(11):1095–1106

Seema HM, Chandrakanth MG, Nagaraj N (2008) Equity in distribution of benefits from water harvesting and groundwater recharge—an economic study in Sujala watershed project in Karnataka. Proceedings of the IWMI-TATA Water Policy Research Program, 2:720–747. http://www.toenre.com/downloads/2008_Sujala_watershed_Seema_MGC_IWMI_TATA.pdf. Accessed 27 June 2015

Seema HM, Srikanthmurthy PS, Chandrakanth MG, Nagaraj N, Veeraraju PV, Muniyappa MC, Dave S (2009) Economic Contribution Watershed Program to groundwater recharge in India, Department of Agricultural Economics, UAS, Bangalore 560065. http://www.toenre.com/downloads/2009_Eco_contribution_of_watershed_prog_IAAE_seema_MGC.pdf. Accessed 27 June 2015

Chapter 16
Economics of Sharing Irrigation Water

16.1 Subdivision and Fragmentation of Holdings

Subdivision of holdings leads to reduction in the size of holdings, while fragmentation leads to increase in the number of holdings. However, a groundwater borewell cannot be physically subdivided. The groundwater-irrigated area can, however, be subdivided among heirs and the senior most heir usually inherits the land where the well is located. Other siblings may/may not be compensated for foregoing the well to the senior. It is precisely due to possession utility that the land gets subdivided. However, availability of groundwater is at stake since it does not depend upon the land size, but is largely a function of degree of weathering underneath. If the farmer inheriting the well with landed property makes up his mind to share the well water with his brothers/sisters, the mushrooming of wells and the associated negative externalities can be reduced as well as the associated costs of groundwater enhancing the net returns.

16.2 Groundwater Friendly Crop Pattern

The results of the study from the central dry zone of Karnataka (Patil 2014) by drawing a sample of 30 farmers who were sharing their well water among brothers/relatives and comparing with an equal number of farmers who were not sharing well water indicated that the crop pattern was dominated by cereals and vegetables in both types of farmers occupying around 60 % of the area. In addition, farmers who were sharing wells were devoted to flower crops, which are groundwater-friendly. For farmers who shared their well water, the average age of a functioning well was 33 % higher than control farmers. In addition, their proportion of successful borewells was higher (60 %) compared with control farmers (50 %). The amortized cost of drilling and casing per well per functioning borewell was ₹ 5500 in shared well farms, as against ₹ 14,000 in control farmers, a saving of 60 % due to sharing of water. The negative externality per borewell on shared well farms was ₹ 1300, which is

M.G. Chandrakanth, *Water Resource Economics*, DOI 10.1007/978-81-322-2479-2_16

four times lower compared with that of control farmers (₹ 6700 per well). The first lesson of sharing well water is the greatest reduction in negative externality by four times when compared with farmers not sharing their water.

16.3 Costs and Benefits of Sharing Water for Irrigation

When the gross returns per functioning well are compared between farmers who share their well water and control farmers, it is apparent that the difference is more than 50 %. This in itself is prima facie evidence for the benefits of sharing well water, which in turn will reduce the negative externality due to cumulative interference. Accordingly, the negative externality is 67 % lower for wells/farmers sharing their well water compared with those who do not share (Table 16.1). The net returns per rupee of cost of irrigation water are 33 % higher for those sharing their well water compared with the control group. The gross returns per functioning well were about 35 % higher for shared well farmers over control, while the net returns were 14 % higher.

16.4 Marginal Productivity of Groundwater Irrigation

Using gross returns per farm as dependent variable (Y Rs), incorporating sharing of irrigation water by families through a dummy variable (1) as against control (0) group of farmers, acre inches of water applied to crops as the independent

Table 16.1 Economics of farms sharing water for irrigation

Features	Sample farmers who shared well water for irrigation	Control farmers (farmers not sharing)	% change
1. Gross returns per acre of gross irrigated area (Rs.)	118,500	112,400	5.15
2. Gross returns per functioning well (Rs.)	761,800	497,900	34.65
3. Gross returns per acre inch of groundwater use (Rs.)	10,600	9000	15.00
4. Gross returns per rupee of irrigation water (Rs.)	28	15	46.43
5. Net returns per functioning well (Rs.)	279,800	240,100	14.20
6. Net returns per rupee of irrigation water (Rs.)	10.83	7.23	33
7. Negative externality (Rs.)	1300	6700	−415
8. Cost per acre inch of groundwater (Rs.)	360	600	−67

Table 16.2 Gross returns per farm regressed on water used

Particulars	Coefficients with significance
Intercept	235,689* (1.97)
Total water used in acre inches (X)	5706** (5.13)
Dummy for institution (D)	211,782 (1.85)
Adjusted R square	0.42
F statistic	16.98**, n=47 farms

Figures in the parentheses are 't' values; Functional form: $Y = \beta_0 + \beta_1 X + \beta_2 D + \varepsilon$; Dummy variable: 1 for shared well farmers, 0 for control farms

variable (X), the marginal productivity of groundwater irrigation is estimated in a linear regression model. The estimated gross return function is Y=235,689+5706 X+211,782D. The gross return at the mean level of groundwater use (72 acre inches) on the control farm situation is Y=235,689+5706 (72)=₹ 646,521 per farm. The estimated gross returns per farm on shared well farmers at mean level of groundwater use is Y=235,689+5706 (88.75)+211,782=₹ 953,879 per farm. The estimated marginal productivity of groundwater is ₹ 5705 per acre inch. The informal sharing of irrigation well water among siblings shifted the gross returns per farm by ₹ 211,782 per farm. The gross return on farms which shared groundwater among siblings was higher by 48% over the control farmers (Table 16.2).

16.5 Transaction Costs

The transaction costs of bringing family members together is low in the case of shared well farmers since a line of control is exercised by the father (or, rarely, grandfather). Virtually, the transaction costs are zero. However, similar efforts, if made to convince farmers to share their well water with neighbours, would involve transaction costs as they do not belong to a family and such experiments are also rare.

In conclusion, families which share well water are rare and the proportion of farmers sharing their well water among siblings does not exceed 10%. The returns from sharing are substantial, exceeding 14% with reduced negative externality in well irrigation by fourfold compared with control farmers. The cost per acre inch of groundwater is lower by around 70% for farmers who share their well water, which in itself provides ample economic evidence of viability of the institution of sharing well water towards sustainable use. The reduction of negative externality due to sharing of water for irrigation to the tune of 415 percent in itself is the prima facie indicator of the magnitude of benefits involved in sharing well water. This resulted in reduction of the cost of water by 67 percent, is another prima facie indicator of the social benefits of sharing well water even among siblings. If this concept can be extended among farmers, it can bring excellent welfare effects. The policy support should be exclusively extended to those farmers willing to share their well water for low water—high value crops so that the sustainable groundwater use is restored.

Reference

Patil KKR (2014) Economics of coping mechanisms in groundwater irrigation: role of markets, institutions and technologies. Unpublished PhD thesis, submitted to the Department of Agricultural Economics, University of Agricultural Sciences, Bangalore

Index

A
Agricultural purpose, 70
  groundwater extraction for, 115
Agroclimatic zones, 111, 185, 198
Alluvial areas, 83
Amortization, 158
  investment, 47, 116
Amortized cost, 20, 21, 47, 52
  borewell, 101
Annual cost of irrigation, 100, 103, 116, 158
  externality defined, 104
  groundwater irrigation, cost of, 104
Average product (AP), 87, 90
  case of, 91
  extreme values, 88

B
Bargaining power
  buyer, 170
  seller, 170, 171

C
Canal command, 184
Capacity building, 20, 81, 193, 195, 199
Carbon footprint
  groundwater, 17
Ceteris paribus, 92, 93
Common property resource, 15, 37, 48
Compensation, 27
Compounded cost, 98
Confined aquifer, 96
Conservation, 193
  soil and water, 199, 201
Consumption function, 25, 96
Consumptive use, 8, 198
Control variable, 173
Conventional irrigation, 23, 132
  groundwater use in, 103
Coping mechanisms, 26, 32, 34, 35, 76
Costing
  irrigation, groundwater in, 100
  irrigation water, 97
Credit supply, 121
Cumulative interference, 3, 11, 15, 26, 31, 38, 109, 110, 112
  irrigation wells, 115
  PRA, 32

D
De facto, 3, 16
De jure, 3, 14, 15, 16, 113
Demand function, 93
  sand, 48, 66, 68, 69
Demand side, 18
  emphasis on, 193
  groundwater, 100
  management, 191
Desiltation, 121, 172
Discriminant function, 136
Diviners, 32, 33, 35, 38
Division of labour, 90
Drip irrigation, 17, 75, 103, 122, 125, 134
  adaption of, 137
  case of, 23
  investment behavior of, 135
Drought year, 141, 142, 144, 147, 148

E
Ecological cost, 12
Economic efficiency
  water use of, 126, 132
Economic life, 100
  borewell, 100, 104
  of borewell, definition of, 100
Economic profile
  farmers, 159, 160

M.G. Chandrakanth, *Water Resource Economics,* DOI 10.1007/978-81-322-2479-2

Economic scarcity, 97
  groundwater, 120, 126, 137
Effective demand, 3
  vegetables and flowers, 115
Efficiency, 12
  economic, 126, 132
  groundwater markets and, 150
  water use, 21, 23
Elasticity of production, 91
Elasticity of substitution, 95
Electrical meter, 18, 116, 194
Environmental cost, 12
Equity
  implications, 35
  issues, 26, 30
Externality, 15, 25, 37, 41, 44
  defined, 104
  explanation fo, 27
  intertemporal, 114
  negative, perspective of, 28
  positive, perspective of, 28, 30
  reciprocal, 21, 25, 27, 37, 47, 96, 99
  unidirectional, 37, 46

F
Farm management, 91, 107
Filter point wells, 46, 48, 51–53, 58, 77
  failure of, 47, 48
Food security, 1, 6, 7
Fugitive, 13, 14, 113, 114

G
Geographical area, 6, 7, 125
Governance, 16
  role of, 11
Green revolution, 85, 99
Gross cropped area (GCA), 6, 49, 53, 126
Groundwater markets (GWM), 120
  and efficiency, 150
  farmers
    economic profile of, 159
    income, sources of, 162
  institutional arrangements in, 168
  recent development, 149

H
Hard rock areas (HRA), 2, 7, 17, 25, 42, 101
High interference, 111, 121
Hydrogeology, 96

I
Illegal quarrying, 70
Indispensability, 95
Inflection point, 88, 90
Initial failure, 34, 99, 113
Institutional area, 21, 26, 98, 110
Institutions, 16, 96, 143
Intercept dummy, 134
Intergenerational inequity, 173
Intertemporal externality, 26, 114, 116
Intricate, 11, 13, 42, 96, 113
Invisibility, 95
Irreversibility, 95
Irrigable command area, 173
Irrigation intensity, 100
Irrigation pumpsets, 103
  diesel, 17, 18
  in India, 9
  power supply, 19
Isolation distance, 11
  policies of, 121
  violation of, 104

J
Jevons paradox, 17

L
Law of diminishing marginal returns (LDMR), 90, 92
Law of diminishing marginal utility (LDMU), 92
Low interference, 112, 121

M
Marginal cost, 6, 12, 115
Marginal external cost (MEC), 28, 46
Marginal farmers, 28, 44, 159
Marginal private cost (MPC), 28, 44
Marginal productivity, 89, 90
  of MP, 134
  of water, 172
Marginal productivity of water, 90, 137, 172
Marginal product (MP), 87, 88, 89, 90, 91
Marginal returns, 11, 182
Marketable surplus, 120
Micro irrigation, 17, 96, 125
Myopic, 4, 78, 114, 180
  extraction, 180, 181, 182, 183, 184, 189

N
Nash equilibrium
  groundwater niche market, 171
  groundwater pricing, 169
Natural recharge, 83
Negative binomial probability, 14, 113

Negative externality, 21, 27, 41, 42
  perspective of, 27
  unidirectional, 44
Net present value of benefits, 189
Net returns per acre, 49, 52, 53, 55
Net returns per acre inch, 136
Net returns per rupee, 58
Niche market, 149, 168
Non agricultural purpose, 170

O
Observed reality, 90
Open access, 14, 113
  resource, 15
Opportunity cost, 4, 64, 182
Optimal control theory, 182, 189
Optimum output, 89
Overdraft, 7, 42, 109
Overextraction, 28, 36, 44

P
Partial budgeting, 202
Partial derivative, 92, 93
Payback period, 99
Permanent income hypothesis, 110
Physical scarcity, 97, 99
Pigouvian tax, 46
Plumb bob, 33, 34
Pollution, 96
Positive externality, 25, 132
  perspective of, 29
Poverty
  social benefit, 23
Power pricing, 121
Power tariff, 20, 120
Premature failure, 14, 15, 34, 47, 109, 194
  concept of, 99
Price mechanism, 96
PRIVATE, 39, 40
Private investment, 3, 8, 112, 194
Private property resource, 48
Production economics (PE), 87, 91
  definition of, 92
Production function, 27, 47, 88, 90, 93
Productivity, 1, 6, 89, 192
  of water, 89, 90, 137
Property rights, 3, 6, 11, 13, 14, 16, 110, 113
  definition of, 172
Pseudo illusionary returns, 66
Public good, 8
Pumping lift, 114, 117, 181

R
Radio isotopes, 81
Random variable(s), 173
Recharge, 2, 11, 14, 16, 36, 43, 75, 111, 173, 181, 182, 183, 202
  of groundwater, 47, 56, 58, 65
Recharge coefficient, 180
Reciprocal externality, 15, 21, 25, 26, 37, 47, 182
Relative income hypothesis, 110
Renewable, 110
Resilience, 1, 6, 95
Return flow coefficient, 176
Riparian areas, 43, 44, 47, 58, 65
Riverine ecosystem, 199

S
Sand excavation, 47
Sand mining, 43, 69, 75, 79
  analysis of, 65
  and NSMAs, 58
  effect of, 43, 44
  influence of, 48
Semi confined aquifer, 3
Silk filatures, 150
Slope dummy, 134
Small farmers, 30, 159, 171
Social capital (SC), 16, 23
Social optimal extraction, 46
Spongy layer, 47, 76
Sprinkler irrigation, 8, 84, 116, 126
State variable, 173
  costate, 173
Stock and flow resource, 95
Storage structure, 99, 100, 116, 120
Storativity coefficient, 176
Subdivision and fragmentation of holdings, 99
Submersible pump, 116
Subsidies, 85, 121
Subsistence life, 99, 100, 104
Sunken cost, 47
Supply side, 13, 16, 18, 192, 195, 201
Surface water, 2, 3, 8, 9, 97, 194
  physical properties of, 10
Sustainability, 11, 198
  addition of, 12
  of groundwater, 14, 172
Sustainable extraction, 1, 4
  of groundwater, 12, 99

T
Tank command, 189
Tank rehabilitation, 23, 172, 201

Technical efficiency, 126, 132
Topography, 3, 169, 199
Total product, 87, 88
Total product (TP), 89
Transaction cost (TC), 4, 14, 16, 104, 199

U
Unconfined aquifer, 13, 96, 113
Unemployment, 23
Unidirectional externality, 15, 25, 37, 47
Urbanization, 43
Utopian, 90

V
Valuation, 95, 96, 199
Variable cost, 47, 78, 96, 100

W
Water buyers, 171
  economy, 167
Water harvesting, 139
Water rate, 8, 97, 198
Water sellers, 159, 162
Watershed area, 139
Watershed institutions, 142, 144
Watershed programmes, 139
Welfare loss, 25, 28, 29, 96
Well log, 114
Willingness to pay, 135
  estimation of, 42
Winner's curse, 30, 35

MIX
Papier aus verantwortungsvollen Quellen
Paper from responsible sources
FSC® C105338
FSC
www.fsc.org

If you have any concerns about our products,
you can contact us on
ProductSafety@springernature.com

In case Publisher is established outside the EU,
the EU authorized representative is:
Springer Nature Customer Service Center GmbH
Europaplatz 3, 69115 Heidelberg, Germany

Printed by Libri Plureos GmbH
in Hamburg, Germany